Catapult

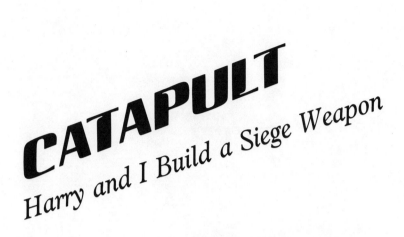

CATAPULT
Harry and I Build a Siege Weapon

Jim Paul

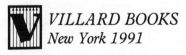

VILLARD BOOKS
New York 1991

Grateful acknowledgment is made to New Holland Publishing, London, for permission to reprint an illustration by Leonardo da Vinci
from *Il Codice Atlantico* as it appeared in *The Crossbow* by Ralph
Payne-Galway.

Library of Congress Cataloging-in-Publication Data

Paul, Jim
    Catapult: Harry and I build a siege weapon/Jim Paul.
      p. cm.
    ISBN 0-394-58507-0
    1. Catapult.   2. Paul, Jim.   3. Engines of war—
California—San Francisco Region.   I. Title.
    U875.P38   1991
    623.4'41'097946—dc20   90-50664

Book Design by Charlotte Staub
Manufactured in the United States of America
9 8 7 6 5 4 3 2
First Edition

ACKNOWLEDGMENTS

Thanks to the director, staff and board at the Headlands Center for the Arts—who made this book possible—and to the others sketched herein, particularly Harry.

I need to acknowledge the following written sources for the chapters of historical narrative: Henry Gough's *Itinerary of Edward the First Throughout His Reign*; Paul Johnson's *A History of the Jews*; Emanuel Raymond Lewis's *Seacoast Fortifications of the United States*; Sir Ralph Payne-Gallwey's *The Crossbow*; Richard Rhodes's *The Making of the Atomic Bomb*. In addition, I would like to thank the following interview sources: Colonel Milton C. Halsey in the Headlands; Antonino Lo Monaco in Siracusa; Archie Wilson at Stirling Castle; Chaim Shapiro at the *Jerusalem Post*. Also of great help to this project were the Government Tourist Boards of Great Britain and Italy, the Exploratorium in San Francisco, and the British School at Rome.

Thanks as well to Chuck Verrill, Gordon Lish, Peter Gethers, Stephanie Long, Holly Blake, and the others who helped this book along. Apologies to the National Park Service.

—Jim Paul
July 1990, San Francisco

Contents

Catapult

Chapter

CHAPTER 1

Red Creek Quartzite

It had occurred to me that holding an old rock might be like looking at the stars. So when I found myself on a writing assignment in Vernal, Utah, I decided to take a side trip into the mountains to hold the oldest rock I could find. I rented a red LeBaron and drove up 191 into the Uintas, which are ancient mountains, older than the Rockies.

Early in the ascent, I passed through the Age of the Dinosaurs, a place where, eighty years before, a scout for Andrew Carnegie had come looking for dinosaur bones. Carnegie had just built a new lobby for his museum back East and wanted a dinosaur skeleton "big as a barn" for it. The scout found the spine of a huge diplodocus cropping out of the earth outside of Vernal, in a layer of greenish rock called the Morrison Formation, and later hundreds of creatures were unearthed here, the victims, it seemed, of some prehistoric flood.

These dinosaurs had brought me to Utah. I was there to write an article about Dinosaur National Monument for *The Washington Post*, and I'd spent several days in Vernal, researching those creatures and feeling more and more displaced among their ancient bones. The earth had no real need for human beings, they seemed to say.

This feeling partook of the general underlying sentiment

around Vernal. The town has adopted a dinosaur motif for commercial purposes, advertising creatures anthropomorphized and humorous, a stegosaurus in a bib and bonnet, a tyrannosaurus rex in sunglasses and a bikini, and on the front of the Chamber of Commerce pamphlet, an apatosaurus in a straw hat with flowers on the brim, carrying a book and a parasol. To me these Late Jurassic cartoons looked like a defense, in that isolated town, against the immensities, the onslaught of so much country, range over range just out the back window, and of so much strange past lying like the dinosaurs within it.

In the exhibit-crammed fieldhouse in Vernal, I'd watched a couple show their daughter the models of the mastodon, bigger than a modern elephant, and of eohippus, smaller than the modern horse. To her daughter, the mother had added matter-of-factly, "When Jesus came, he made all the animals the right size."

Climbing the Uintas in my LeBaron, I thought about this mother, and wondered about her need to favor the ages with a face. It was an understandable impulse, I thought, akin to my own. I hoped an ancient rock would resolve things, would get me beyond this sense that gigantic alien forces, big reptiles among them, had always possessed the earth. The rock might let me feel some intimate connection to the immensities of the past. The night sky had allowed me that once—in western Ireland, when I had looked into the celestial coral reef of the Milky Way and felt welcomed there. So I drove into the mountains and hoped.

The state of Utah has put up friendly blue-and-white signs on the roadside to explain the various immensities around Route 191, calling it the Drive Thru the Ages. CURTIS FORMATION says one sign, adding—as if stones had homes—HOME OF FOSSILIZED SQUID. The Uintas came up like a bubble from the center of the earth and burst through the layers of stone above them. The upheaval pushed the broken edges of the strata toward the sky, so that today the blacktop north of Vernal cuts through ridges that once lay flat in the earth, and as the road climbs in altitude, it plunges chronologically downward, backward in time, through outcrops and hogbacks, each one of rock older than the last.

My rented LeBaron put millions of years behind us, soon leaving the dinosaurs long in the future. I drove between ridges of pink Navajo sandstone, and through roadcuts in the piled plates of the Chinle. Beyond them, the road wound among the Moenkopi's red powder hills, and finally beneath pocked towers of Park City sandstone, the pinnacles ghostly, as if they had been destroyed once and had risen again into the sunlight to tell about it, as they had.

Then I reached the shale of the range itself, and drove over the crest and down toward the dam at Flaming Gorge. The Uintas thrust through the older crust to a height of about 14,000 feet, and on that northern side have opened very ancient strata to the surface. The road there dives into a deep gorge called Sheep Creek Canyon, and switches back and forth along the canyon wall as it descends into the fault. Beneath the pocked turrets I found red rock, Weber sandstone a billion years old. This rock was Precambrian, I read, a term like postmodern, suggesting that what it names is so mysterious as to require identification by what it isn't.

Beneath that, the stone got crazier. The south wall of the canyon rose as a comb of yellow rock, its face etched with flamelike fissures. Beyond this sentinel, nearly at the bottom of the canyon now, I passed a memorial to a family of campers killed in a flash flood, and found a steep field of red plates, at weird angles to one another: Precambrian crockery. Then, reaching the stream at the bottom of the gorge, I saw a kind of a stone orb, big as an apartment building, that seemed to have pushed its head through the floor of the canyon.

I got out of the car and scrambled up the mound of sharp scree to its face, where I picked up a chunk of the same stone, big as a grapefruit, heavier than a telephone, one side still bubbled by the fire in the earth. It was Red Creek quartzite, a smooth rock, denser than marble, with broad pink and white stripes banding it, a stone two and a half billion years old, half the age of the planet itself and older than life on earth.

I tried to think about a billion, tried to remember that it wasn't twice a million, but a thousand times a million. And I tried to make a million a thousand thousands, tried to make one of all those thousands be that day, and tried to make that day be a year. But I'm not a geologist, and despite my arithmetic, the thing was just a rock in my hand, big as a grapefruit, heavier than a telephone, pink and white. I decided to take it home to San Francisco for the hall table. Though I took no solace from it, it could still be my talisman against the past, like Vernal's apatosaurus with its book and its parasol.

In the airport at Salt Lake, I had to take the rock out of my bag and show it to the woman operating the weapons detector. It was just a rock, I said, not a weapon. As a joke for myself only—because, as a sign warned me, it might have been a crime if I'd said it out loud—I remembered that the rock might indeed be a weapon, though a stupid one to try to hijack anything with.

That thought probably arose because I was angry at the airport in Salt Lake anyway. My trip had nearly been ruined on arrival. Debby, the clerk at Dollar Rent-a-Car, had tried to confiscate my credit card when I tried to rent the car with it. Debby ran my card through a machine and cheerfully informed me that it was invalid. She was kind of cute, though overly made-up and wearing that Farrah Fawcett hairstyle that has somehow remained in fashion in the Deep West—two out-curled waves of frosted blond hair framing her pink face. She made a phone call, during which she watched me fidget behind the counter. She hung up, smiled, and said, "I've been instructed to confiscate your card."

Illustration from *The Crossbow*, by Sir Ralph Payne-Galway, The Holland Press, Ltd., London, 1903.

She had slipped my credit card somewhere I couldn't see it, and I was alarmed by the prospect of losing the thing—I had my story to do, five days until my return flight, and no chance of getting a car without my credit card. As a free-lance writer, I had to lay out these costs as overhead. The *Post* would eventually reimburse me, and I relied on my credit card to carry me over until then. So I had mailed the credit card company a check before I had left home to get my balance safely below my limit for the trip. But that check was evidently still in the mail. I was stuck—again. I tried to remember why I ever started free-lancing. A salary seemed like streets paved with gold.

"It's all a mistake," I said hopelessly to impervious Debby. "Please don't take it."

She was cheerful and adamant. "They have my name and place of employment," she said. "I have to."

Ultimately I called the bank in Seattle that had issued me the card, and had to be horribly ingratiating to get them to talk to Debby and cover for me. When she took the receiver, she said, "Hi, this is Debby at Dollar Rent-a-Car." She listened to them a while and finally uncovered my card, which she had hidden under the stapler. Then she informed me I'd have to lay out $200 to rent the car. "Thanks," I said as I paid up most of the cash I'd brought with me. "You're a nice person, Debby." Finally she handed me back the card.

Getting searched for a weapon at the end of my trip, I remembered pleading and smiling and charming my way through the ordeal with Debby. I was glad to see the last of that place. I put the useless rock back in my carry-on bag and boarded the plane for home.

Back home at my apartment in the Mission District of San Francisco, I called my friend Harry while I was still unpacking. I had taken out the rock and had thrown it on the bed, where it sat in the dent that it made in the afghan and comforter. It looked less old than ever. I picked it up and held it as I called Harry. I told him about Utah and the credit card fiasco. The check was in the mail, I said. Sure, he said. I was abstractedly hefting the rock around, and the image of the stone flung into the air came to mind. Without thinking about it much, I said, "You know what I'd like to do with you, Harry? Build a big catapult and shoot stones into the ocean."

The idea was a whim, a surprise from some dark corner of my mind. I knew it would appeal to Harry, but I didn't expect him to take it as seriously as he did. If he had dismissed it at that moment, nothing would have come of it. But he treated it as merely unfeasible. "That'd be fun," he said. "But I don't have time to do something like that.

"Besides," he added, "you never make stuff. If we ever made anything together, I'd do all the work."

"I make stuff," I said. What had I made lately? I thought. A paragraph, a sandwich.

"You have no idea about making something like a catapult," Harry said. "It would be technical. If you wanted to do it right, you couldn't just throw it together in a day."

"It wouldn't have to be that big a deal," I said. "Just something we could take into the Headlands and shoot."

"I don't have time," said Harry. "I have a family. Besides, who would pay for it?"

"It wouldn't have to cost that much, Harry," I said. "We could build it from scraps."

"From scraps," said Harry derisively. "Look, just forget about it, OK? Believe me, it would cost money. Money is the first step in anything technical."

Ross wanted to talk to me, Harry said. The ten-year-old boy had been in the background, asking about the dinosaurs, until Harry finally put him on. I told him about the Carnegie scout and the big bone. They had a little park in Vernal, I told him, with life-sized model dinosaurs. Could you get up on them? Ross wanted to know. "No," I said. "They were just for viewing."

"Put Harry back on, would you, Ross?" I said.

"Harry," I said, when he got back on. "What if I got the money for it?"

"Forget about it," Harry said.

"What if I got the money?" I asked.

"You just do that, Jim," he said. "You just write a polite note to the Defense Department. And then you let me know what they say, OK?"

He laughed and hung up.

I got off the phone, surprised by the force of this whim to build a catapult. Where was that coming from? I wondered. I had never wanted anything like it. I had never even owned a weapon. I held on to the rock and wandered around the apartment, agitated, fancying that I felt some insistence from the stone itself, from the Red Creek quartzite. Its heft was just out of my range, its bulk too much for my arm to manage, had I decided to throw it. So the thing could have been calling for some sort of machine to toss it harder than I could. It was an odd thought, that a stone could suggest anything, as if some rocks ask with their whole weight to be put to flight.

Besides, I thought, we had to build this catapult for Harry, if for no other reason. He'd been nervous about money ever since his new baby was born, conscious of his responsibilities, reminded by this beautiful baby, Julia, of his duty. She had her way of doing that. From birth she had possessed the most piercing shriek, as well as the wits to employ it. Her scream made my eardrums rattle like bad speakers. No one could do anything in the presence of such a scream except attend to it. So Susan was nursing at all hours, and Harry was working his day-job full tilt, frantic with money terror, lacquering furniture as if it were a sporting event. He needed this catapult, I thought, as a therapeutic measure of irresponsibility.

Besides that, I was irritated and challenged by the way Harry had ridiculed my ability to get the money. I knew how to get money—I had worked as a grant writer, and had gotten money for other people's projects. If I could somehow find the money for a catapult, Harry would have to eat his words, a tantalizing prospect.

But these reasons, I knew even as I thought them, were secondary—if not entirely spurious—hypotheses in the face of the curious potency that this idea of making and shooting a catapult had for me. Even calling it an obsession didn't diminish it. Maybe the idea had had me, instead of vice versa. I needed the catapult, and I couldn't quite say why, and that interested me. Maybe I wanted to be a boy again, on the grand scale that a man might manage. I couldn't know unless we went ahead. I paced around my apartment thinking about it, shifting the Red Creek quartzite from hand to hand, until I laughed out loud and thought I knew where I could get the money.

The next evening my girlfriend, Sara, and I took a salad over to Harry and Susan's. At the time they were living on the third floor of a light-industrial warehouse in West Oakland. The building was a big, pink block with a water tower on the roof, and Harry had met Susan there when they'd each taken live-in painting studios in the place, among several woodshops, a small factory that made gift-shop ceramic bears, a bicycle-wear manufacturer. After Harry and Susan had seen each other for a couple of months, Harry moved in with Susan and her son, Ross. They got married at City Hall a few months later.

Ross was the same age as Harry's son, Isaac, who lived with them on weekends and over the summer. The boys got along. The family had lived in the studio for a couple of years when Julia was born, and this beautiful child had so disturbed the neighbors with her scream—in a warehouse, yet—that one night somebody called the cops. Harry had to stand there and convince these two big women cops that nothing was wrong, that the baby just screamed like that. So before this catapult idea got started, Harry and Susan had been house-hunting on the weekends, looking for a better place for a family, a place Julia could scream to her heart's content. Mortgage, I thought, that's what the catapult is up against.

Susan buzzed Sara and me in that evening, and we climbed the three flights of stairs. The warehouse was great. It was sunny and spacious, the outer walls mostly windows, and the hallways inside broad and wooden, a labyrinth where we skateboarded with the boys after hours, rocketing down the wooden ramps and around the corners, finally out into the parking lot. Before the baby was born, Harry liked to skateboard. He liked to fall down a lot, to barrel off the loading dock.

Susan opened the door. She had her hair back, and she had been rolling chicken in flour—her hands were white with the stuff. Inside the kids were carousing, as usual, the boys wrestling and chasing and the baby everywhere, shrieking occasionally. Above the ample loft, the big space between the roof beams and the tops of the walls was full of horizontal summer light. I left Sara with Susan in the kitchen, and went back into Harry's studio to find him.

Burly and bald, Harry sat in his rocker, staring at a painting, at a few marks like the chinks and holes in a wall, the rest still blank. Staring was the main work of painting, he told me once. I never looked much at his work while he was doing it, though, because it always changed. His work had its way with him, I used to say.

During his first marriage, Harry decided to take up kyudo—formal Japanese archery. He read everything on the subject and got quite good at it. He'd made his own primitive bows. But he didn't take everything about the sport seriously, which was his downfall. One was supposed to enter the range in a certain posture and attitude, fold one's robe with a certain posture and attitude, breathe properly, say the right things, concentrate in some perfect egoless way, not think about hitting the bull's-eye, and then hit the bull's eye. All Harry did was hit the bull's-eye. On more than one occasion he enraged his teachers with his sloppiness and his irreverence, not to mention his sure aim. For a time, kyudo replaced drawing as Harry's obsession. He was unsure about drawing, about its ultimate value. But after a couple of years of shooting arrows, after he got divorced, he gave up kyudo. He hung the big Japanese bow on the wall of his bedroom, and went back to drawing, though he still kept his big archery book in his studio.

I asked him what our catapult would look like, if we ever made one.

"Like a crossbow," he said. "That's what catapults actually were—big crossbows. Those other kind—the Monty Python kind—they were called something else."

"What did they look like?" I asked.

"I know what you're trying to do," he said. But he showed me a picture of a crossbow, anyway, a woodcut in the big archery book. He pointed out some of its features.

"Could you shoot rocks with it?" I asked.

"Oh, yeah," Harry said. "They did that. You'd have to modify the bowstring somehow to hold them."

I'd rather make the Monty Python kind, I said. Harry began to warm to his subject then, trying to convince me that this crossbow catapult was the best. The other kind broke down every other shot, he said. They were stupid. They were from the Dark Ages.

"You just want to make another bow," I said.

"These were the best kind," Harry insisted. "What I want has nothing to do with it."

"Look, Harry," I said. "I'm serious about this. Would you help me do this, if I came up with some money for it?"

"You couldn't afford my time," Harry said.

It would be a weekend project, I said. It would be fun. How much would we need, just for supplies? I asked.

"I don't know," he said. "Five hundred bucks maybe."

"It couldn't be," I said.

"Believe it," said Harry.

"What if I got it?" I demanded.

I guess Harry felt like he could afford to look like a sport, and get me off his back at the same time. "OK, Jim here it is," he said. "If you get the money, I'll help you build a catapult." He felt safe, I could tell. I'd never get the money in a million years, he was thinking.

When we came out of the studio, the women were setting the table beneath one of Susan's big paintings, a cartoon of a pair of California clowns enjoying the good life by a pool. I told the women about our plans. "We're going to build a catapult," I said.

"*If* Jim can find some money," said Harry, adding for good measure, "which I doubt he can."

"Why would you want to build a catapult?" said Sara.

"We just do," I said. "It'd be fun."

She had a tactful suggestion. "Couldn't you guys just make a garden or something?" she said.

Susan was more direct. She just beat her chest with her floury fists and howled.

CHAPTER 2

Headlands

I've always been a rock-thrower. I could throw rocks into water all day. One summer I lived by myself in a cabin on a lake in Upper Michigan, morainal country, and the beach was just shingle, acres of flat smooth stones, some of them disks of archaic coral. In my solitude, I looked to be absorbed by something, and I went down to the shore every day at sunset, and threw rocks for an hour or so, until it got dark. I was absorbed; there was some controlled and comforting fury in it. That summer I didn't want to skip stones, as I had done before in that place—I wanted to throw them high and far. I liked the way the flat rocks turned over in the air in the middle of an arc, then entered the water nearly splashless with a sound like puncture. Every day I threw rocks until I couldn't see, until my right arm throbbed. Over the summer, it grew more muscular than my left. Sometimes I'd get an hour's rock-throwing in before breakfast, as well. I must have moved tons of beach offshore that summer.

I had thrown rocks with Harry on a few occasions, often without thinking, outdoors somewhere while we talked about other things, pitching stones unconsciously as if by some elaborate, effective twitch. Once in the Marin Headlands we had become purposeful about our rock-throwing. We had gone out there for a hike, and had ended up on a bluff a hundred feet above the ocean, where we noticed a yellow plastic sand bucket floating in the water, rising and falling in the swell beyond the surf line. Without saying anything we began throwing rocks at the bucket, at first casually, in the midst of our talking, and then more intensely. Gradually the bombardment of the bucket became the only business of the hour. It was in part a contest, of course, and neither of us managed to hit the bucket for a long time—though we splashed it and moved it around with near misses. Finally, after an hour, when we both at last began to grow weary of the game, I picked up a big rock, big as a softball, and with a roar heaved it off the cliff. As if it had eyes, the rock clobbered the yellow bucket, submerging it for a time. Sheer luck. Harry had to try a few more rocks after that, but he couldn't hit the bucket. Finally we hiked back to the car and drove home.

I was thinking about bombarding the yellow bucket when I realized where I could get the money. I knew I wanted to fire the catapult off of a deserted military battery up there in the Marin Headlands. The thing and the place required each other. I imagined us on a bluff over the ocean where the army had left a concrete emplacement, where they had put up thick walls as if to cradle the thing and had cut away the terrain to get a wider view of the water. It would be a place where our stone-thrower could command the sea.

That thought collided with another, and I came up with the Headlands Center for the Arts. Suddenly I knew how the building of the catapult might proceed. I would examine the question itself. I would observe this impulse to build a catapult. I would call it art and apply for a grant.

So on Monday I called JD, the director of the Headlands Center for the Arts, and told her I had a project I wanted to discuss. We made an appointment, and one day later that week I drove up there, across the Golden Gate Bridge and over the hills into the fortress of the Headlands.

* * *

That place had always had some powerful attraction for Harry and me, and we'd been all over the huge and craggy tract. We'd been down the faces of the cliffs and into the sea caves where at high tide the waves battered the stone. Sometimes, from the cliffs above, we watched huge waves, twenty or thirty feet high, wheeling and colliding out over Potato Patch Shoal, a mile beyond the point. These waves came from the northwest, down from the Aleutians, and when they broke into the sea cliff, we could sometimes feel the shudder through the rocks.

As I drove up to see JD, I came north across the bridge, and saw the Headlands rise on the seaward side. They were California hills the breakers had sheered open, the faces of the sea cliffs a tangle of red stone over green. The green rock—basalt—was older, poured molten out of a rift in the sea floor, the red chert cast over it later in seasonal sheets, in layers of tiny sea creatures that had died, fallen to the bottom, and turned to stone. The formation is too much, even for geologists. Mashed and cast ashore and lifted into mountains and cut away into cliffs, it's like some crazy dialect. The geologists call it Franciscan Melange.

Behind these cliffs, the Headlands is a big empty place, ridges exposed to the wind off the ocean. Only the army and a few hardy Portuguese ranchers ever lived here. The Indians left just shell mounds in the coves of the Headlands—they lived over the hill, where Sausalito would be. But the exposure was just what the army sought, for it offered points of command above the Golden Gate. The army reserved the Headlands for its guns, closing it to civilians for more than a century. But now even the army is gone, having abandoned that desolate place in the mid-seventies. They stripped the facilities and left them to ruin.

These army ruins Harry and I had found particularly irresistible, and we had explored many of them, having figured out a trick for finding secret places, gun batteries and ammo rooms. No big trees ever grew in the Headlands, just scrub on the seaward ridges and thickets of little willows in the valleys below, Sausalitoes with their roots in the streambeds. But the army planted cypress and eucalyptus—big trees—around the gun batteries, to provide windbreak and perhaps—by dictation from some army manual—to camouflage the guns. So Harry and I would climb the ridges and head for the big trees. Beneath them, we'd usually find air vents and chimneys protruding from the slope, or wide concrete arches, sometimes with bars across their entrances or steel doors welded shut. If we could get in, we often found the tunnels streaked with graffiti. In one of the bunkers, above a deep stagnant pool in which hundreds of black salamanders wriggled, someone had made some additions in spray paint to an official notice. "This Ain't OFF LIMITS Anymore," it read.

Harry and I had hiked up to the missile radar site, a place on a peak called 88C, and had climbed around this ghost facility with its concrete monoliths and empty platforms raised to the sky. Beneath one trapdoor, we'd found a small lookout post dug high into the hill, the sea's horizon in the slit across the western wall, and from a hole in the floor, an eruption of raw cable.

We'd gone through the empty ammo tunnels between the batteries without even a match, Harry completely night-blind and walking behind for once, as we felt our way down the stone corridors. The blackness down there was total, so dark that my eyes seemed to project odd effects like light. Sometimes it seemed as if we walked beneath a starry sky, though the reinforced concrete ceiling was just two feet overhead. At other times, some ghostly light seemed to be shining on my back, casting a long pale shadow ahead into the blackness of the tunnel.

Exploring the ruins was sometimes dangerous, often frightening. Plus there were rangers in the Headlands—the place was still federal property—and we always had to watch it, figuring that at the very least they would throw us out if they caught us. There was a possibility we'd get lectured and ticketed, and have to pay some federal fine, and the rangers had power to arrest, as well. All that added to the allure of the Headlands, of course.

Physically, the art center in the Headlands was a legacy of the army. When the army left, the National Park Service took over the place, and found itself overwhelmed by the army-sized facility. In buildings alone, it had barracks, warehouses, a chapel, a movie house, a gymnasium, and several large homes up in the cypresses above the lagoon, officers' quarters. For several years, all these were simply boarded up. Then the Park Service invited other groups to reoccupy the buildings, and one of these groups became the Headlands Center for the Arts.

One August, I watched a group of children from the center's art camp take over the big gun battery on this hill, the summit of the Headlands. They put on a performance for their parents in the former gun emplacement. In the deep concrete bowl where a million-ton cannon and its counterweight had swiveled, the children lay down sheets of white cloth, covering themselves with it, as their camp counselors buried them in several dozen big white balloons. Their parents peered over the lip of this concrete nest, and the children seemed to awaken within it to the sound of tambourines. Then, dressed in wings and tatters to appear like butterflies, they climbed up ropes out of the gun emplacement, knocking loose the balloons. A strong wind blew off the ocean through the tunnels and corridors in the hollow hilltop, and it caught these disrupted balloons, blowing them down the corridors, through the tunnels, and out the other side of the hill. The children danced after them, spinning around and shouting in the reverberating tunnel, leaving their parents behind. I drove past these bunkers as I climbed the ridge. The kids would be a good precedent to mention to JD, I thought.

Mainly the art center at the Headlands was dedicated to art as exploration, not to the picture, as it were, but to the act of picturing. When the center took possession of the old barracks at Fort Barry, its board commissioned David Ireland to do the first art project there, but Ireland didn't put paint on anything—he took paint off. He and his workers spent a year arresting the decay in the abandoned barracks, stripping many layers of army paint off the walls, and revealing the pressed-tin ceilings, the webs of cracks behind the plaster, the wood grain of the balustrade. What they found they sealed with finish or wax, as a display. In the attic, they unearthed photographs of the servicemen, their duty rosters, the cots on which they'd slept. They reopened the mess hall and the latrine. Sometimes people came into the art center, saw Ireland's work, and asked where the art was. This was fine with Ireland. "I like the piece not to have a flag on it that says it's a work of art," he said.

I had decided to cast the catapult in this light as well. Building and firing it, I would propose to JD, would be a way of reliving the thought and action of the army men who had lived out there in the Headlands, for one thing. We would be observing the process of arming, I would say, recapitulating the development of weapons technology, putting on the mask of the weapon-maker. It would be a Conceptual Reconstruction. It might not even appear to be art. Harry would never buy any of this, but maybe he didn't have to know.

The sea claimed more and more of the horizon as I crested the ridge and tried to think of the catapult entirely in this pure and theoretical light. But occurring in the midst of these thoughts was another, like a frog-call from the lagoon of my unconscious. It would be fun. It would be so much fun.

*   *   *

I pulled up in the gravel lot between the big white barracks buildings, and went in through the Ireland lobby. I found JD in her office. She was a tall, sophisticated woman who liked to wear textured hose, and though we'd been friends for some time, she could still manage a sort of brusque New England formality in the office that made me nervous.

"What's on your mind?" she said. Most of my rehearsal went out the window when I spoke. "I have this friend, Harry," I said. "He and I would like to build a big catapult and shoot rocks into the ocean with it."

JD thought a moment and finally said, "Why?"

"To observe the impulse," I said.

"To observe the impulse to shoot a catapult?" she said.

"Yes," I said. I gathered my wits a minute and tried out my catapult-as-art theory on JD. I said that Harry and I would be engaging in catapult consciousness. "It'll be Conceptual Reconstruction," I said.

JD laughed out loud. She laughed that first moment she heard about the catapult, and she kept laughing for several months whenever I talked to her about it. I think she enjoyed what she thought was my sense of irony about the project. Whatever it was, she kept laughing until she actually saw the catapult.

Still, she was taken with the idea. She treated it as a joke, but thought it might make an interesting exploration anyway, and asked what she could do to help.

"We could use one of the gun batteries," I said.

She said we'd have to check it out with the Park Service. My spirits fell when I heard that. The rangers had let artists do some odd things in the Headlands. One had danced head-first down ropes strung from the army trees; one had brought in beehives and turned the army gym into a huge honeybee colony. But I had a feeling the rangers might draw the line at bringing in a siege weapon.

JD picked up the phone. As she punched in the call, I

wondered idly if the Park Service really had any idea of what they were getting into when they invited artists into the Headlands. In the old days there were people like Protestants and pipe fitters. Now it seemed like everyone was some kind of artist.

JD was in great form. She smiled into the phone and said, "Charlie, I have a couple of artists here who want to build a catapult. That's right, a catapult."

After what seemed like a long moment, she said, "Well, what they want to do is bring it up here, take it up on one of the batteries, and"—she paused for effect—"give all the rangers free vacations to Hawaii on it." I could hear the tiny cawing of the ranger's laughter on the other end. She had him on the hook, I realized.

She listened for an even longer moment, then said, "That's right, as a kind of prop." I didn't like the sound of that, I thought. JD went into a string of affirmatives: "Yes, right, OK, fine, great then." Then she hung up. "He says you have to get a permit," she said.

"Oh," I said.

"But he doesn't think it will be a problem," she said.

"Fantastic, that's great, thanks," I said, adding, "Do you think the center would look at a proposal to fund this project, at maybe five hundred dollars, for parts?"

"I'm sure the board would look at it," JD said. After a moment, she added, "There'd have to be some kind of public component."

"What do you mean?" I said.

"You'd have to present your findings," she said. "In a public lecture or something."

Findings. Harry's going to flip when he hears about findings.

"No problem," I said.

JD gave me a deadline, a few weeks away, when I should have the proposal ready if I wanted it considered at the August meeting of the board. I said, "Great," a couple of times and got out of there, quickly and delicately, so as not to upset the deal, which to me seemed to stand balanced for the moment, precarious and amazing.

I drove home and called Harry. "We got it," I said. "Got what?" he said. I explained to him that I had gone up to the Headlands to see if I could get the money for the catapult. "You went out there?" he said, incredulous.

I tried to make the whole thing sound as matter-of-fact as possible. I tried to explain to him about the catapult as an artistic project. I told him we were going to investigate catapult consciousness.

"And they bought that?" Harry said.

"The rangers will give us a permit," I said, glossing a little. Harry didn't really need to know all the details, I thought. "And the board will entertain a proposal about the money."

"In other words, we don't got it," said Harry.

"Yet," I said.

"We don't got it," said Harry. "And I can't believe you think that they would ever let us do something like this."

As usual Harry's naysaying just made me seem sunnier and sunnier. "They might," I said. I remembered JD saying there would have to be some kind of public component. But I didn't want to mention findings and public lectures just then. As it was, I felt stretched pretty thin between what I had to say to Harry and what I had to say to get the money. I just said, "Everyone was very nice."

"I bet they were," said Harry. "They probably thought you had a catapult out in the yard, loaded and aimed."

That made me laugh.

CHAPTER 3

A Country of Miracles

Men have been shooting things off the cliffs of the Marin Headlands for a long time, either to drive away their enemies or just to answer the challenge of the broad water somehow. Arnold Palmer hit a golf ball for a commercial here, socking one into the sea for the cameras. It was supposed to be a funny ad, suggesting that Palmer could drive a ball all the way across the Golden Gate, which is a mile wide in that place. In reality, although Palmer must have kept his head down and let the club do the work, as he recommends in his book, entitled *Hit It Hard!*, his championship drive must have looked puny enough in the scale of those cliffs. The tiny white ball must have traced a brief arc out beyond the lip of the bluff, a thread of a gesture toward the sea, and then have disappeared into the chasm.

Palmer hit that ball off Battery Kirby, a concrete pad and housing originally built as an emplacement for two twelve-inch Endicott cannons, powerful guns. One forgets how completely sufficient the weaponry of the last century was. Twelve-inch cannons throw projectiles that are twelve inches wide, and these cannons at Battery Kirby, installed in 1895 and named for the secretary of war, threw those twelve-inch, one-thousand-pound projectiles for eight miles. At the Headlands

I saw an army photo of one of those old cannons being fired at night. The fireball from the gun's muzzle had overexposed the film on one side of the picture. On the other, blurred by the vibration of the explosion, the four gunners stood isolated from each other and turned away, seeming both despondent and devotional as they shielded their faces from the blast.

By the 1840s, even before California joined the Union, men had placed cannons on these cliffs lining the strait. The Mexicans put up a fort on the southern shore and aimed eight green bronze guns through gaps in its walls. Cast in Peru in the seventeenth century, they bore among other decorative flourishes the coat of arms of the king of Spain, and are now trophies of war. Today, two of them flank the entrance to the Officers' Club at Fort Scott in the Presidio.

When the Americans took that Mexican fort, the Castillo de San Joaquin, they cut away the cliff to sea level, built a squat brick fortress at the water line and called it Fort Point. In it, they placed 126 cannons, four tiers of the big smooth-bore muzzle-loaders known as Rodmans poking from square embrasures in the walls. The Rodmans fired at a low angle, so that their cannonballs could skip across the surface of the water. During the Civil War, the Confederate raider *Shenandoah* anchored off the Farallon Islands, thirty miles beyond the Gate, with orders to shell San Francisco and hold it hostage. The captain must have imagined scores of these 400-pound red-hot cannonballs bounding in a flock toward the sides of his wooden vessel. The *Shenandoah* weighed anchor and headed north.

Fort Point now crouches beneath the Golden Gate Bridge, under its own arch at the southern anchorage, still appearing impregnable, but long obsolete. The only such stone fort on the West Coast, it is in one sense a throwback to the medieval castle, which also depended on vertical stone walls for its defense. The invention of the cannon in the fourteenth century made such land forts obsolete, but as seacoast defenses they hung on. Though a cannonball might spall the brick to a depth of an inch or so, guns fired from the rolling deck of a ship could not hit the same spot on the fort's wall over and over, an action necessary for breaching it. But in the accelerated development of weaponry that accompanied the Civil War, the special technology of rifling—grooving—the inside of cannon barrels became standard practice, and this seemingly slight innovation instantly made masonry coastal forts useless. Rifled cannons could throw spinning, pointed projectiles, and could throw them harder and more accurately than previous cannons. These shells could penetrate brick.

Union gunners made practical application of this new technology at Fort Pulaski, downriver from Savannah, during the Civil War. Having dragged the heavy rifled cannons through the swamps of nearby Hilton Head Island, the gunners set them up and shelled Fort Pulaski quite neatly, crushing a point on the wall opposite the fort's powder magazine. As soon as daylight opened between the Union cannons and the Confederate powder, the rebels surrendered, and this easy capture of a brick fort by rifled guns came as a shock to military tacticians. Despite their obsolescence, these masonry defenses continued for decades to be funded and fortified—by armor-plating the brick, for instance, a trivial enterprise, preliminary to new thinking.

The Fort Point cannons, like all the generations of armament in the Headlands, were never tested in battle. The army waited for more than a century in a place that seemed logical, for an enemy who never arrived. The enemies changed over the years, of course. The first guns here waited for the Spanish, then for the British, then for the Americans. For a time, Argentinians seemed to be a threat, and early on the Russians established trading posts on the horizon, at the Farallons. In our own century, after the Germans and Japanese, the enemy was Russian again. But the guns on the Headlands were never called into action. No big gun on the whole West Coast, in fact, was ever fired in anger. There was only one opportunity to fire them.

On June 25, 1942, at 2:00 A.M., a Japanese submarine surfaced off Fort Stevens, Washington, and lobbed about two dozen shells against the hillside, near big guns like those at the Headlands. The Americans, a National Guard unit, didn't return fire. The commander claimed the submarine with its three-inch cannon was out of range of his ten-inch gun, and besides, he said, he didn't want to give away his position. It had been a good life, up until then, there in Fort Stevens. He and his men would spend the rest of the war giving their positions away in places like Guadalcanal.

Though they didn't often attack the mainland so brazenly during World War II, the Japanese did routinely lay siege to the United States by incendiary balloon. They sent thousands of balloons aloft into the jet stream, their wicker baskets loaded with hair-trigger explosives. News of this Wizard-of-Oz kind of warfare stayed out of the press for the duration. Floating to America with the weather, some of these balloons came to earth as far east as Milwaukee. Only three people were ever killed by them, though, a family of picnickers who found one of the balloons hanging from a tree in the Great Northwest, and who were overzealous in their attempts to find out what was in its basket. Dorothy and her little dog, maybe. They hadn't read anything about exploding balloons in the newspapers, of course.

During this period, the Pacific Ocean didn't seem wide enough. Japan seemed to be just over the horizon from Ocean Beach in San Francisco. The army built massive concrete casemates to contain the stupendous guns that would fend off the Japanese when they came. They put these bunkers mostly underground, digging into the high ridges and pouring walls and ceilings of concrete thirteen feet thick. The single sixteen-inch cannon in each bunker was sixty or seventy feet long and weighed more than a million pounds. The barrel stuck out from beneath a round beetle-browed stone canopy, and could loft a one-ton shell about thirty miles. In practice firings of these sixteen-inchers, their projectiles pierced twenty feet of reinforced concrete before exploding.

For all their power, though, these big fixed guns were probably obsolete on the drawing board. They were sitting ducks for airplanes, not to mention missiles, which were soon to replace cannons as the cannon had replaced the catapult. And within ten years of their installation, having never been fired in anger, the big guns were taken down and sold as high-grade scrap to the appliance industry, steel for the Speedqueens and Frigidaires and Toastmasters that became so popular after the war. Some of the steel was cut into razor blades.

So in the fifties and sixties, the guns in the Headlands gave way to missiles. The army installed a covey of "birds," as the servicemen called them, in a hollow designated as Site 88L, behind the sea cliffs. Battery Bravo, they named the installation. The missiles were Nikes—named for the Greek goddess of victory—and were surface-to-air missiles, stored underground and lifted to their firing racks at the surface by a massive, howling hydraulic elevator.

The first Nike—the Ajax—wasn't large. At twenty feet, one might have balanced on the luggage rack of a station wagon, like an oversized surfboard. Douglas Aircraft put out a press release on its product, describing the Nike-Ajax in advertising copy as a "formidable guardian of the free world" and "dart-like, with delta-shaped cruciform fins." The Ajax had three conventional warheads containing fragmentation bombs that would throw shrapnel through enemy airplanes. They might as well have been defending against arrows and burning pitch. The army put them in the Headlands in 1954, and took them out in 1958, when it became clear that enemy airplanes were no longer a threat, and that these missiles wouldn't be able to shoot down other missiles.

The army put in a new Nike in the Headlands, a fatter one

called the Hercules. It sat on the bundle of four missiles that composed its solid-fuel booster, and had a nuclear warhead with a switch that could adjust the device's explosive power from the equivalent of fifty tons of TNT to that of two thousand tons of the explosive, the size of Little Boy, the Hiroshima bomb. An article in *Army Magazine* announced the Hercules matter-of-factly, as if it would soon be put to use, the only nod to the unprecedented scale of the new weapon's power in its reassurance that the warhead "will be employed at altitudes where the effect of blast heat and radiation on the ground will be negligible." The manufacturer designed the Hercules to fly to a height of about twenty-eight miles, above enemy formations, then to dive down amid their ranks and explode, emptying the atmosphere from horizon to horizon.

Each Hercules had a red shield on its side bearing an inscription that informed anyone who could read Latin that possession is a matter of successful defense. Needless to say, these missiles were never fired, though they were readied for use during the Cuban Missile Crisis in 1962. Generally the nuclear warheads in the Headlands were kept unloaded—their plutonium cores were elsewhere, perhaps across the bay at the Concord Naval Weapons Station, to be ferried in by helicopter at red alert. Then a crewman would place the radioactive cores inside the lenses of explosives in the warheads, making the bomb capable of releasing the vast energies locked in every particle of matter.

Unlike all previous generations of weapons, the Nike-Hercules missiles in the Headlands were not replaced by newer, more powerful armament. The missiles were outlawed in 1972, when Richard Nixon and Leonid Brezhnev signed an international treaty in Moscow—the ABM Treaty. After thirty years of nuclear threat, these two cold warriors put down these arms, at least. The treaty all but banned defensive missiles like the Nike-Hercules, which were obsolete in any case, considering the size and might of the forces they opposed. The whole world was within range of the new weapons, everywhere vulnerable, anywhere a potential battlefield, and so seacoast defenses no longer made any sense. So the positions in the Headlands were bargained away. Missiles like the Nikes remained at only four places on earth—in rings around the urban capitals of these two nations and around intercontinental ballistic missile sites in their desolate interiors.

So in 1976, the army held a ceremony in the Headlands, recited the lineage and accomplishments of its units, and left. The order to leave didn't mention the ABM Treaty. It said that the department had concluded that the program should be dealt a lower priority status, and therefore it authorized inactivation. The army abandoned land within sight of downtown San Francisco that, but for the bunkers and the bridge, looks about the same as it did when the Miwoks gathered mussels in the coves, before the Europeans drove westward across the whole continent, as if to take this last point, and finally, oddly, to abandon it. Today park rangers maintain two last unloaded Hercules missiles in the Fort Barry bunkers as historic objects, like George Washington's hatchet.

And at the state dinner for the successful ABM negotiators, given at the American Embassy in Moscow, Nixon and Brezhnev congratulated each other on their work, and Nixon served dessert. It was Baked Alaska, which Nixon explained was "hot ice cream." Brezhnev turned his great, impassive, beetle-browed face to his table of diplomats. "America," he said, "is a country of miracles." Chief among those miracles, perhaps, was that America at that point still existed at all.

CHAPTER 4

Mock Rocks

The Bay Bridge is all business—a massive main artery, ten lanes of airborne freeway on two decks. The following Saturday morning I had blasted over it twice already, and it was only ten o'clock. I had to convince Harry to go to the library with me, and he finally agreed only if I'd drive. So I picked up the Prince at his house, and drove back to San Francisco to go to the library's main branch in Civic Center, a place lifelong-resident Harry had never been.

On the way over there, and as if casually, I slipped JD's public component past Harry. I didn't need him to be flipping out just then, when he was barely on board. "By the way," I said, "if we get this money, I'm going to have to do some kind of public lecture about the project."

It slipped past him. "That's your territory, Jim," he said. "If we get to shoot it, that'll be enough for me." He added, "I'm sure you'll have a great time with your audience." What he didn't realize then was that I would have to talk about him in the lecture. But he didn't think of that, only of how awful public speaking was. "For me," he said, "it would be torture."

I could tell Harry still didn't believe we would get the money, and that he felt he didn't have to listen closely when I mentioned the details of the bargain I'd struck. Findings, I thought again. I knew it wasn't a word he would like.

He loved it in the library, though. He ran his hands over the marble stairwell, exclaimed at the reading room, and by the time we got back into the science and technology stacks, he was too impatient to take the books back to the tables. So we sat on the floor between the shelves and paged through articles on catapults and crossbows, and books on weaponry and the lore of arms.

The first big book, called simply *Weapons*, was a compendium, like a *Norton Anthology of Horrible Things*. It was a coffee-table book, handsomely pictured, which began with clubs and spears and ended with hydrogen bombs. In between were objects of absurd cruelty and specialization—impaling stakes for use in concealed pits, for instance. Who would display such a book on his coffee table? I thought. Just looking at it made me feel a little queasy. I hadn't quite classed our catapult with these other weapons.

Harry pointed out the decoration on some of the early weapons, especially those intended for hand use. They bore elaborate carvings, inlaid ebony and mother-of-pearl, and other features seemingly more suited to jewelry than to deadly instruments. A German wheel lock pistol from about 1580, for dueling no doubt, had its butt inset with gilt strips and carved ivory, decorated with birds and masks. The bulbous end of the butt bore cameos of beautiful long-haired women, and the trigger had been delicately turned, as if on a tiny lathe. The craftsman had carved a ram, a crab, and several men's faces into the elaborate hammer.

Maybe, I thought, these decorated weapons had simply been produced for a luxurious class that could afford to make all objects delightful to the eye. Or maybe these designs were the echo of magical icons, placed on a weapon in ancient times to enhance its power—in the Dark Ages in England it was supposed that certain runes, carved on a stick, would transform it into a magical and lethal wand. Probably, though, the decoration was just an attempt to glamorize murder, to provide an excuse in advance for the deed committed with the weapon, as if killing didn't count if it was done tastefully with a beautiful gun.

As the centuries proceeded, I noticed, it seemed to become less and less necessary to decorate a weapon, especially a big weapon. The bazooka seemed required to do nothing but function; in fact, only the grooves on the grip of its trigger gave any indication that it had been designed in relation to a human body. And of course the latest and most powerful weapons, hydrogen bombs, were invisible inside the sleek missiles that delivered them. I imagined simply black spheres, abstractions with no characteristics at all, until they went off.

Maybe we ought to play with this idea, I thought, and decorate the catapult somehow, perhaps in the faux marble that Harry could do in his shop.

"That would be stupid," Harry said. "Why don't we just make it pink and put little bows on it?" Making anything except a plain functioning device would expose us to the worst kind of ridicule, he stated.

Right in the middle of this coffee-table weaponry book, where the technology first went beyond hand-to-hand combat, were siege engines, ballistae, and catapults. A stone thrower was a ballista, I learned. Catapults usually shot bolts. The word meant shield-piercer in Greek. It was the Greeks, under Philip of Macedon in the third century B.C., who developed the first very powerful torsion-spring catapults, and systematized their use. Philip also hired Aristotle to be his son's tutor, and the sum of these gifts was empire itself. That son used Aristotle's systematic thought and these great engines of war to conquer the known world as Alexander the Great. Alexander's catapults used animal sinews, the book said, woven efficiently into elastic ropes, as their source of power. Just as Harry had said, these catapults looked like, and in fact were the ancestors of crossbows, smooth-torqued in their action and exact in their aim.

Harry liked an illustration he found in the book, the reproduction of an old engraving of the siege of Jerusalem by the Romans in 70 A.D. In the foreground were several catapults, in the background the city wall, topped with battlements, above that the temple. The catapults looked like old-fashioned biplanes to me, manned by centurions with those whisk-broom plumes on their helmets.

"Torsion springs," Harry said. He pointed out the bundles of sinew that propelled the stones—the things that had looked to me like the wings of biplanes. The picture excited Harry, and he plunged ahead into the trove of books that we had piled around us on the floor.

In a bound volume of the journal *Scientific American*, we found an issue with a cover story devoted to ancient catapults. We both exclaimed when we saw the cover illustration, a detailed drawing of the same kind of Roman torsion-spring catapult as those pictured in the illustration of the siege of Jerusalem. Another illustration accompanying the article by Werner Soedel and Vernard Foley was even better. It was the representation of a huge and precise machine, maybe twenty-five feet high, built of timbers thicker than roof beams. The arms of the bow projected from two twisted bundles of sinew—like Popsicle sticks inserted in twisted rubber bands—and these arms were strung with a bowstring which in turn was cocked by a ratcheted winch at the base of the sloping stock.

"Look how big it is," Harry said. Beside the massive device stood the outline of a Roman soldier, to show the scale of the machine. He was barely taller than the big feet of the catapult's pedestal.

This picture was just about all Harry needed to see. I had to prevent him from tearing it out of the magazine. We took the volume to a Xerox machine and got a copy. In the months ahead, I was to be amazed at how much Harry understood just by looking at that picture. He really did not need any further description of the mechanism in words. As for me, it took me some time just looking at the drawing before I could find the projectile, a stone about as big as the centurion's head, which lay cradled in a net woven into the bowstring. I was amazed when I finally saw the stone—it was so tiny compared to the mechanism. Mechanical power was apparent, I understood, visible the way grandfather clock weights are. If you needed immense mechanical power, you needed an immense machine. I thought about that, taking Harry home across the Bay Bridge, as Harry got the picture out again to look at it. I was glad for his enthusiasm. We were in for a bigger job than I had imagined. I was going to need Harry.

Over the next few weeks, I wrote our proposal. I collected catapult stories from friends. A stage designer named Eric told me that his company had once built a catapult for Ozzy Osbourne. Osbourne had recently broken up with his heavy metal group, Black Sabbath, and wanted to enliven the finale for his first solo performances by catapulting raw meat into his audience. So along with fog machines and lights and scrims, the staging company designed and built a device that would heave handfuls of chicken livers for fifty or a hundred feet. The techs came up with a catapult powered by bungee cords, which they dubbed the Ozzy Osbourne Liver Launcher. Until they figured it out, Eric said, the crew had a hard time firing the Liver Launcher during the concerts. Their catapult proved quite powerful, but it tended to hold its sticky ammo too long before releasing it, and until the crew could achieve the proper loft, the thing merely spattered chicken guts into the backs of the necks of the security guards lining the stage.

Another friend, Tom, said that he'd helped build a similar device for a fraternity at college. A huge contraption powered by yards of surgical tubing, the thing threw fruit at a neighboring fraternity, and required eight or ten guys to hold the tubing, and another eight or ten to pull it back. The force of the device made mush out of the softer fruits and vegetables, virtually pureeing tomatoes, for instance, and crushing the shells of watermelons in the thrust of the launch. Tom said that the fraternity brothers experimented with various kinds of food before they found the cantaloupe, a perfect projectile for their purposes, which they could shoot for blocks. Throwing fruit, I guess, made the whole thing a joke, even though the cantaloupe probably struck with a good solid wallop, certainly enough to smash a window or even crack a wall. The police, in any case, didn't find it funny, and eventually put an end to their fruit wars.

Harry's friend Dusty told me another catapult story that ended up with someone being arrested. Some guy up near Sacramento, he said, was being driven mad by low-flying planes. The guy was a baker and a day-sleeper. He'd work all night, and then come home to be robbed of his sleep by the roar of these planes buzzing his house. He phoned the airport to complain, but was told that his house was not beneath any approach path. Frustrated and furious, the guy built a big arm-and-sling catapult, the Monty Python kind, in his backyard. When he was finished, he brought buckets of bread dough home from work. He'd get home at dawn and load the catapult. When the planes came, he'd fire lumps of bread dough at them. Dusty claimed the guy hit a couple of the planes, scaring the pilots. It was serious, said Dusty. If a gooney bird can bring down an airplane, he said, no doubt a lump of bread dough could do so too. Eventually one of the pilots found a wad of dough stuck to the belly of his Cessna and went to the police, who began cruising the neighborhood where the baker lived and eventually caught him. He got three years probation, Dusty said.

And about this time I went to a party at the house of another friend, a literary critic named Bobby. The guy was brilliant, no doubt, but when he didn't like a piece of work, he declined no tactic, including ad hominem arguments, by which to dismantle his opponent. He had set his own novel in prehistoric times, among a society of cave dwellers like those at Lascaux.

I had never been to his house before, and he showed me around. Bobby lived in a big loft South-of-Market, and slept eight feet off the floor, on a platform above what passed for his kitchen, accessible only by means of a thick rope tied to a rafter. He and whoever slept with him, he said, had to climb that rope. He showed me his collection of bones, and I told him about the catapult.

He said he knew a guy who'd built something like that once, he said, from a kit he'd ordered out of *Soldier of Fortune* magazine. The thing shot big bolts. The guy had taken it to Africa and killed an elephant with it, he said, just to prove he could kill the world's largest animal with a weapon of his own making. The bolt went right through the elephant's neck, he said. I didn't really believe Bobby's story. Blue whales are the largest animals on earth, I said.

Then he told me he had something I might want to see, and we went into a back room, where he got out his Uzi. It was a cheap, black, boxy, sawed-off looking gun, both disgusting and fascinating. He held it in both hands and said he needed it in his neighborhood. This is South-of-Market, I said, not the Gaza Strip. Then he offered to let me hold the gun, and I declined, trying not to appear uncool about it. I really did not want to touch that gun. After that I left the party, in a state of turmoil. Bobby had really blown it with me, I thought. Somehow I had tolerated his professional behavior, even if he'd hurt friends of mine. But the gun was beyond the pale. I drove home agitated.

The next day I told Harry about going to the party at Bobby's, and about my feelings about the Uzi. I was glad to be investigating this catapult thing, I said. Seeing that gun close up, I recognized that a person who showed the barbarity of the culture at large seemed disturbed, even sick. All my life my country has been building huge, sick weapons in my name, I said.

"That sounds exactly like something someone would say at a party," said Harry. Harry couldn't stand parties, he added, for just that reason.

All this went into the proposal, which I finished on the night before the deadline. I related the catapult stories, and mentioned that the Park Service had in some measure already approved our plan. I included the illustration from *Scientific American*, hoping it would have the same effect on the board as it had had on Harry. And my agitation at seeing the Uzi at Bobby's helped me write about the catapult in a cool and distant, even ironical way. I didn't claim it would be fun. I presented the catapult as a sort of prop in a demonstration. I wrote that Harry and I intended "to rehearse consciously the impulse to make and use this weapon, and in the process, to render that impulse less automatic and less powerful." I asked for the $500 and figured that the project would take us a month.

The board met one evening in July, and JD called me the next day. They had voted unanimously, in a discussion marked by laughter, to give us the money. When I hung up, I called Harry up right away, pulling him off the shop floor at work. "We got the money," I said.

Harry was flabbergasted. All he could do was sputter that he couldn't believe it. Then, a couple of times, as it kept dawning on him, he said, "This means we have to do it." And finally, he insisted there had to be some hitch. "No hitch," I said, rubbing in my victory.

But there was a hitch, it turned out, and a bigger one than simply our having to present our findings in a lecture. JD called me back later to say that when she had informed the rangers of the board's decision, they had come up with a proviso to their permission. There was a rule against bringing weapons into a national park, it turned out. We could bring the catapult into the Headlands only if it was a piece of art, and not if it was a weapon.

"What would be the difference?" I asked.

"A weapon shoots real rocks," said JD. "Art shoots mock rocks."

"What are mock rocks?" I asked.

We'd have to decide that, JD said. Maybe papier mâché. "I didn't think it mattered," she said. "A Conceptual Reconstruction doesn't have to shoot real rocks, does it?"

"I guess not," I said. I said I had to talk to Harry—I'd get back to her. When I hung up, I got the chunk of Red Creek quartzite off the hall table. It had a decidedly unmock look. Then I called Harry. He was furious at the whole idea. "What are we? Going to do all this work and then shoot foam rubber out of the thing?"

The more we talked about it, the more irritated we got. Couched as a question of safety, this no-real-rocks requirement seemed to sum up the whole liability of being an artist—its ineffectuality. I remembered the decorated dueling pistol in the coffee-table book, the beautiful long-haired women engraved on its butt. The maker of that sixteenth-century pistol had decorated it to give its deadly function the blessing of art. But mock rocks would ensure that our catapult remained aesthetic and artificial. The gesture might be brilliant, the statement exact, the execution perfect, but all of it would seem as light as those kids' white balloons, blowing through the battery in the wind, pretty and fanciful, but risking nothing. Besides, the army had been up there shooting things for years, we agreed.

Harry was infuriated, incredulous, at being forced to choose between making art and producing a machine that worked. I tried to joke with him about it. Maybe a device that threw bread on the water might be a strong statement, I said. Maybe we could ask the rangers if we could throw those round sourdough loaves. But Harry wouldn't even joke. He hated the idea of strong statements, he said. He declared that he would have nothing to do with any such artsy-fartsy bullshit. The catapult would work, he said, it would look the way it had to look to work, and it would throw real stones, or he was off the project. No pink paint, no ribbons, no mock rocks. He was adamant about it.

What could I say? We'd worry about it later.

For Harry, I summoned my bravado. "Well, look," I said. "Right now we've got the money, and I think we should just build what we want to build. We'll just go ahead and make a machine that can shoot real rocks. Big ones. Heavy ones. Wall-smashers."

"Damn right," said Harry.

But even then I knew that shooting real rocks would come down to defying JD and the National Park Service, and it took some effort to ignore the question. I tried rather lamely to frame the issue conceptually, in artistic terms. Magritte, I remembered, could play with the artificiality of his painted image by labeling his picture of a pipe with the words "This is not a pipe." And conceptual artists could put a frame around anything, a urinal or a checkbook register, for instance, and declare it art. So what was mockness in the first place? If it was simply a degree of artificiality, then what could be more artificial than my imported piece of Red Creek quartzite, invested as it was with my own particular meanings?

In the final analysis, I told Harry, we could simply declare the stone to be mock. This is not a rock, we could say. And we were the artists, after all.

"Come off it," Harry said, adding, "You should have been a lawyer."

So I called JD and just said that everything was OK. She said she'd process our check. And later that week I had a dream in which Harry and I stole a car for a joyride. We drove the wrong way down the freeway, firing an automatic pistol out the window. The cops gave chase, and we abandoned the vehicle, running into a building, a school, in which we found the locker room. The gun gets thrown under a bench—we don't want to get caught with it—and Harry and I take off in opposite directions, to make our capture more difficult. When I see him later in the dream, Harry smiles and asks me ironically if I've lost anything. Then he hands me back the gun. Shocked, I recognize it as my own.

CHAPTER 5
The Model

When we made our model, I recalled Veronica, who had introduced me to Harry and Susan on a double date, when we'd eaten blackened redfish at the Elite Café. Veronica was a red-haired sculptor who lived in the studio next door to Harry's and Susan's, in the big pink warehouse. At that time she spent her off-hours recomposing pieces of old furniture into glaring little shrines, often painted red and studded with spikes—golf tees, actually, glued on upside down.

Veronica made models at work. She'd worked on the Death Star for *Star Wars*. She had made a model of a Hopi village for a museum display, and someone from the movie had called to offer her a job after having seen it. Veronica worked on several movies in the model shop. She appeared briefly in a TV documentary on the outfit, modeling a realistic mineshaft down which a doll-sized Harrison Ford could plummet in a tiny handcar with his tiny friends. At the shop, Veronica came to specialize in organic effects. She could make a redwood forest on a tabletop as a place for the rocket ships to land. Veronica called the shop Boys Town.

She took me to a showing of one of the worst movies she worked on. Veronica didn't pretend that it was going to be any good. She'd seen the dailies, after all. But she wanted to see her work in front of a real audience, and so we went. Near the end of the movie, as so often happens, the hero is trying to get his helpless friends out of a cave. It's a vast cave, like the Grand Canyon underground, and before he can get out of there, he comes to a place where the ground just falls away, and it's miles and miles deep. The only depth the film had, said Veronica. This chasm was her effect. At the bottom of the chasm was water, and soon they were sailing a ship on that water, which was an underground lake. Then there was an earthquake, during which huge boulders fell into the water. Those were hers, too, she whispered in the dark. But these effects were quick. They just flashed across the screen, and they had taken many months to make.

Between films, Veronica made her sculptures, and when she finally went East to be an artist in New York, she gave me one of them, a red podium with a bandaged cane for one leg and jaws lined with golf tees. She gave it to me because I thought it was funny, which, before I lived with it, it was. The thing stayed around a long time, and I finally found it anything but humorous, and moved it into the closet, where it glowered from the dimness whenever I opened the door.

When Harry and I began to build our project, I went to that closet to see if I had anything that I could make a catapult out of, and found Veronica's podium-with-fangs. By this time, the thing seemed positively sentient, gaping among the empty boxes, gnarly and powerful. What was I up to? it seemed to demand. What did I want? I thought about trying to saw the thing up into catapult parts, but it didn't seem worth the effort, so I just shut the door on it again.

Like any ideal, the Roman-style catapult we'd seen pictured in *Scientific American*—powerful, rectilinear, precise, and so huge—seemed out of our reach. So while we waited for the check in the mail, we argued—usually amid the ruins of dinner—about what our catapult was actually going to look like. I put forth some rather festive conceptions for the thing, mostly to bother Harry. His least favorite of my designs was a catapult made of a clutch of sailboard masts, bound together somehow at the ends, and bowed outward like a skeletal pumpkin. The stone would sit in a nest at the top of this pumpkin, and when the tension was somehow released, the stays would unspring, shooting the stone straight up. I imagined the whole thing festooned with silk streamers, even the stones. When he heard about this fiberglass pumpkin catapult, Harry made a great show of deriding my mechanical sense. For one thing, he said, such a catapult would drop the stones back down on top of us. This was something I had not foreseen. I had to admit that my interest had always been in getting the stones into the air—not in where they came down. Such a poet, Harry said.

But even Harry thought the Roman model was beyond our capacities—not to mention our budget. Bundles of animal sinews might be hard to come by. Harry's own dinner-table designs tended to be heavily fortified. He was the mechanic. He was the archer. He had some idea of the forces involved. He made a drawing on a napkin that looked like a boxcar built of railroad ties, with the shooting mechanism deep in the middle somewhere. Seeing the drawing, I observed that at some point we might want to transport the thing without the use of a heavy crane. I could see us trying to get that boxcar up some steep dirt road in the Headlands, the two of us toiling away beneath Harry's tremendous creation, like a scene out of *Spartacus*.

We both knew that these debates were mere formalities, mostly bluffs, a way to get started on the project without really getting started. We had until early December to build and fire the catapult, and that seemed ages away. We didn't actually know what we were going to do. I don't think Harry really let himself believe that the check was coming. So it wasn't until late summer, when the money finally arrived, that we did any-

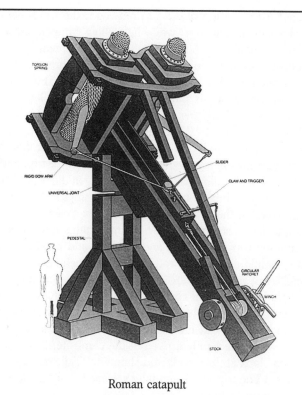

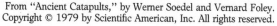

Roman catapult

thing concrete. Even then, we didn't start on the catapult proper. We decided to get some more rocks.

My mother arrived. She'd come out from her home in the East to go to summer camp in Yosemite, a two-week elderhostel. My mother is a lively, funny woman called Mabe by her friends. Now in her late sixties, she still beats me at tennis. She returned from this trip with a snapshot in which she appeared astride a gigantic fallen redwood, the trunk perhaps twenty feet in diameter, while her fellow campers clustered below on the forest floor, gazing up in amazement.

I drove her up to Yosemite, which is about three hours from San Francisco. My mother has a bunch of Thurberesque expressions—if she is piqued, she may shout "Good Night, Nurse." When I told her about the catapult project, she said, "Criminy," adding, "That sounds like something a pair of little kids would do." So I changed the subject after that. Still, it was nice having my mother all to myself on the trip, and we took in the sights along the way, remarking at the peach groves in the valley, pointing out the raw outcroppings in the foothills. She gasped at the plunge into the gorge of Yosemite.

Harry had also gone up into the mountains that week—on a camping trip with Susan and the kids and another family, Harry's childhood buddy and his wife and their kids. The older daughter of this other family was very demanding and rarely denied. She was about six, and still sleeping between her parents every night. She had her father buffaloed. The little girl's demands had cowed him, and he appeased her at almost all costs. When the family arrived at the campsite, this little girl took one look at the woods and shouted, "Take me home this instant." The trip was eventually salvaged, but the week was full of tantrums beside the waterfalls.

Elsewhere in the woods, I was undergoing this odd reversal, dropping my mother off at summer camp. We found her tent among the rows of them under the big pines on the floor of the valley at Camp Curry, and I made sure she'd brought everything, sneakers and binoculars and bug repellant. She was up there early. Nobody else seemed to be in the big camp, and I left her there unpacking her things onto her cot. I felt a little sad—worse, I'm sure, than if things had been reversed and she was dropping me off in Yosemite—and I drove off down the valley on my own, beneath the peaks and towering walls of the canyon.

The road winds down the valley floor along the Merced

River, and I drove a couple of miles downstream before I pulled off the road at a familiar footbridge. I got out and crossed the Merced by this little bridge, and followed the footpath on up the slope through the ponderosa pines. It led through the woods to the base of the monumental granite wall called El Capitan. So enormous that climbers spend days antlike on its face, sleeping in hammocks pinned to the granite with thousands of feet of empty air beneath them, El Capitan is too large to comprehend without real concentration. Most tourists claim—defending themselves from the implication of the scale of that huge rock—that it is about a third of its actual height. To stand beneath it, beneath the pines that fringe its base like the merest lichen, is to gaze across a vertical world of stone and feel minuscule and harmless.

The hillside at the cliff's base is composed of chips, blocks, and boulders that have broken off the monolith for ages. They were salt-and-pepper granite, young relative to the venerable Red Creek quartzite, and delicate—the white grains sandy and pure, the black ones tiny bubbles of glass. These were the offspring of El Capitan, colossus of the stone world, I thought. I could not resist. These stones wanted the catapult, and the catapult wanted them. I gathered about a dozen big chunks of the granite, stowed them furtively in my pack, took them back to the car, and drove them out of the valley.

Meanwhile, in another part of the forest, Harry had found rocks of his own. He brought home a bunch of them, the same kind of granite, which he'd taken from a stream. The water had darkened and smoothed their surfaces. When he got back to Oakland, I brought my rocks over to his house and we combined our stones and piled them into a cardboard box. The box bulged. It looked like a considerable stock of ammo.

That day in Harry's studio we made two springs out of folded and coiled paper, taped them to pencils, and tied a bowstring of twine between them, to see if we could shoot anything with such a construction. It was a paper catapult, a little slingshot, slightly resembling the Roman model. When he was done, Harry had to have a flat surface to shoot across—this paper thing wasn't about to loft anything—and we went into the kitchen to use the surface of the table. Susan was making dinner. Harry couldn't get the thing to fire by himself, and he made me hold the pencils while he folded a wad of paper across the bowstring. Susan was boiling a big pot of water for pasta. She looked over her shoulder and said, "Sheesh, you guys." Still, the paper model seemed to work. The wad went all the way across the table.

The next day when I went back to Harry's, he decided on the spot that it was time to make a wooden model, and began rooting around in his studio for scrap wood and parts. When Harry makes anything, he takes a long time to get going, but as soon as he actually initiates the activity, he works like a man possessed. He doesn't have a lot of tools—he just brutes things together furiously and becomes distraught if he is impeded at all. I hovered around the fringe of his frenetic activity, offering suggestions.

He hammered some small boards into a T-shape, then decided that we had to have some clock spring for this wooden model, and made me look in the Yellow Pages under "Springs." I thought this was ridiculous. Surely there would be no business that simply sold springs. There turned out to be several. The closest was on East 14th in Oakland, a funky corridor of bars, tacquerias, evangelical storefront churches, billboards for booze, and check-cashing places.

Harry and I argued on the way over there about how we were going to ask our questions. Harry didn't want to tell anyone what we were actually doing, but just to describe the requirements of the job. I thought that this would get us nowhere, and besides, I wanted to tell people what we were making. I thought they'd be interested. I brought one of the pieces of El Capitan granite, to show them. Besides, I said, we might learn something. Harry finally agreed, but told me to do the talking.

We found the place, SH Springs, a grimy front office with a gray desk and a calendar fronting a narrow shop that ran to the back of the building. For a long time we just stood in the empty office. Nobody appeared. Finally an old woman with long gray hair and a face like a ferret emerged from the darkness.

The woman just stared at me as I held out the rock and began to try to tell her what we wanted. Then she turned and screamed something unintelligible into the shop. She motioned for us to go back there. Once our eyes adjusted, we saw a skinny, wizened man at work, clamping something into a heavy double-bladed machine. After a while, he got it clamped and turned toward us.

The guy looked awful. His cheeks were sunken and scarred with pockmarks. He was emaciated and filthy, blackened with soot and stinking of booze. He could have been seventy; he could have been forty. I felt stupid, face to face with this guy, obviously a hard-luck case, beaten down by a grueling life of work. I just held out the rock and asked how big a spring we would need to throw it, say, three hundred yards. The old guy was quiet a long time, and finally he just said, "Jesus H. Christ."

It was a bad moment. Harry took over. He just asked the guy if we could buy some clock spring. The guy didn't say anything, but pointed to a coil of the stuff on the floor—a curled ribbon of blue steel, yards long. Harry got the guy to snap off ten feet of it, using a press that bit off the metal with its thick jaws. We took the ribbon and tried to pay, but the old guy just waved us off, as if it wasn't worth money to him. Then as we turned to go, he finally spoke.

"That could be real dangerous," he said. "You're going to need a monster spring to throw that rock. And if you break that spring, you'll throw shrapnel everywhere. I wouldn't want

to be within a hundred feet of it when you try it." We thanked
him for his advice and got out of there, carrying our loop of
blue steel. I was glad to be back out in the sunshine, and I
said so.

"Not used to that stuff, eh?" said Harry, as we got back in
the car. Harry hadn't been uncomfortable in the old guy's
grimy shop—after all, he said, he'd worked in such places.
When he got out of school, he had taken a job working in his
father's shop repairing pinball machines. Harry's father,
Benny, ran a pinball and jukebox business in the Mission and
South-of-Market districts of San Francisco, and when Harry
was a boy, his father took him on his rounds, mostly to bars
where these machines were. Sometimes the bars were sleazy
places, where addicts and criminals would hang out—and
Harry would stay right with his father. Sometimes there would
be the Spanish songs called *correos* playing, and a pretty se-
ñorita behind the bar who'd give Harry Cokes and treat him
specially because he was Benny's boy. They'd empty the
change out of the machines. Harry's father had bags of coins
in a safe beneath the floor of the office.

When he got old enough to go to work for his father, Harry
fortified the new pinball machines in the shop, putting hasps
and plating and locks on them. He worked under his father's
foreman, a fat man named Red, who had to learn to like the
boss's son. The pinball machines themselves were like primitive computers, Harry said. They had a big cam that tripped
banks of switches connected to relays, regulating the machine's
response when the steel balls fell into the holes above. Some
of the machines had sexy women painted on them—one called
Wood Nymphs, Harry recalled, had a bevy of beauties in short
tattered skirts and Peter Pan caps on its mirrored backglass.

I had a work life like my father's, too, for a while. My father
worked in Washington for the Kennedy administration, as
something like a deputy under-assistant secretary of the in-

terior. I worked one summer in the department, as well—in the budget office of the Bureau of Land Management. Every morning I rode in from the Virginia suburbs, down Shirley Highway in the steamy heat, with my father and his car pool—three other guys with similar jobs in Washington. All summer in the office, I thought about Walt Whitman working in that place—I was an English major and I found the fact that Whitman was ultimately fired from his sinecure in the Interior Department very comforting. All summer—my first not spent on vacation—I reviewed and summarized reports and checked the addition on columns of numbers and made multicolored pie charts for presentations, and then I wandered the marble hallways, thinking about Whitman. July crawled by, and I was more and more appalled by the lives of the people I worked with, who seemed to have no more enthusiasm for their work than I did for mine. They were in it for the duration, though, not just for the summer.

Harry broke up the clock spring when we got back to the studio, snapping it in his bare hands. Then he punched holes in it with a nail, and made a pair of layered wings for the model, which now looked like an odd, squared-off slingshot. He bound the springs into slots in the wood with picture-hanging wire, which he also used for a bowstring, and stole a piece of the boys' Lego set for a cup to hold the projectile, a marble we'd found. He wouldn't let me anywhere near the model. I shouted corrections at him and tried to grab it. Harry just barked at me, pulled the model away and resumed straining over it. Finally I followed him out of the studio, and up the stairs to the roof of the warehouse.

The toy was pretty tough. The wire bowstring was taut enough to be difficult for Harry to pull back, and its recoil threw the marble a long way across the roof. Harry ran around retrieving the marble after each shot, crunching over the tar and gravel, over other people's ceilings, shouting about how good the model was. I was sure that the bowstring was driving the marble into the nose of the launcher. A bowstring just will not adequately control a stone, I argued, following Harry around on the roof. "You've loved bows your whole life," I said. "You just want to make a bow no matter what."

"You don't know what you're talking about," Harry shouted back. "It's working perfectly." Finally I had to race ahead of him, grab the marble, and hold it captive until Harry would give me the model. Then we each shot it a bunch of times, yelling and chasing the projectile around the vents and skylights on the roof until we lost the marble. Harry insisted that I had shot it off the roof. After that we quit, never actually resolving our argument, though Harry would later carve a trough for the stone to ride in, an attempt to answer my objections. And I had to admit that if the real thing threw its rocks at the same scale, we would have a real weapon on our hands.

At that point Susan and Harry moved to their new place, a big old white house a couple of blocks from the warehouse in a residential neighborhood. The new place was enormous, and was a deal—it would probably have been a hundred thousand dollars more if it hadn't been in West Oakland. Even as it was, Harry and Susan went through agony with the agent and the owner and the loan and the appraisal. Harry acted like it was his one chance ever at a house. It was now or never, Harry said. It was make or break. It took them a couple of weeks just to move their possessions, during which I understood it would not be a good idea to mention the catapult. So I lugged boxes and furniture, and let the subject slide. The new house was better for the project anyway, I thought. It had a big back porch, sided with that corrugated translucent green fiberglass and covered with a roof—the perfect place to build a catapult.

The model nearly got lost in the move. We'd given it to the boys, who'd played with it briefly, found it too difficult to cock, and discarded it. When we packed and moved the studio, I found it on the scrap heap. I took it to my car to take home with me.

"What do you want that for?" said Harry. "We know what it can do." Harry wasn't sentimental about things. The model had served its purpose. "It's no good to us now," he said. I took it home anyway, and put it on the hall table with my rock.

CHAPTER 6

The Engines of Archimedes

Arethusa was a wood nymph, a hunter devoted to Diana, the goddess of the moon, and sworn to chastity. Nonetheless, she fell in love with a mortal man, and he with her, and so Arethusa was presented with a dilemma. She prayed to Diana for help, and the goddess granted her divine rescue. As usual, contact with the gods—even help—brought with it a kind of poetic ruin, ending Arethusa's life as a nymph. Diana transformed her into a spring on a far-off island, where she flows today as a fountain, beautiful and for no man to hold.

For his part, her mortal lover found himself bereft, and prayed to Zeus, who transformed him into a river. And although he is a river in Greece, and she a spring in Sicily, a cup of wine poured into the stream will stain the fountain red.

So goes the story of the freshwater spring that first drew Greek traders to Syracuse. Besides the Arethusa spring, the place had other natural advantages: the city arose on the Ortygia, a small island bridgeable from the mainland at the mouth of a deep harbor. The Greek seafarers gradually began to settle this small island, driving the locals into the hills on the mainland and fortifying the Ortygia with a surrounding seawall, until Syracuse appeared as a single, seaborne castle, the Arethusa spring at its heart.

To the Roman orator Cicero—writing two hundred and fifty years after the downfall of Syracuse—the town still seemed elegant and Greek, "the most beautiful of Greek cities," he said. Though it might have looked Greek to Roman eyes— white overall, a temple to Apollo, the graceful amphitheater like a huge shell set into the hill—the city hadn't been ruled by Greeks for four centuries. In fact, when the former colonists at Syracuse finally crushed the Athenians in 413 B.C., the defeat marked the closing of the Greek period in the world. The Syracusans struck coins to commemorate their independence, and for one historical instant—a century or so, with Greece defeated and Rome not yet risen—the city seemed to stand at the center of the universe. The great power of the Western world paused there before it passed on to Rome, and in the interim, at its zenith, Syracuse was a place of high culture and high technology. Aeschylus witnessed productions of his own plays in the city. And its ruler, a tyrant called Dionysius the Elder, began a research-and-development program to invent new weapons. Historians of technology have credited this program with the invention of the first catapult.

Even without the weapons, this walled island with a source of fresh water within it was virtually impregnable. The rising power of Rome gradually forced the city into an alliance, but Syracuse was blessed during that time with a peace of fifty years under the rule of another tyrant, Hiero the Second. And Hiero had added to the city's imposing natural defenses by commissioning the city's leading scientist, the brilliant geometer Archimedes himself, to build and deploy a seemingly supernatural array of catapults.

Archimedes himself stank. He believed with Plato that the realm of disembodied and abstract thought approached some pure ideal, providing glimpses of the perfect world of which this material realm is but a flickering shadow. Therefore he smelled like a goatherd. He goes on record as one of the first men in Western history to have been so thoroughly an inhabitant of the mental world, in his case the world of abstract geometry, that he neglected to wash. The stink of the man annoyed his patron, Hiero, who had the genius forcibly bathed on occasion. Even under such duress, Plutarch records, Archimedes would lose himself in thought, and would trace geometric figures with his finger in the tub grime.

That he made his most famous discovery in the tub seems to follow, though Hiero gets no official credit. History doesn't record whether Archimedes was being forcibly bathed on the occasion of this discovery—the relation of buoyancy to the displacement of water, published in a treatise entitled *On Floating Bodies*—only that the mathematician himself ran naked through the streets of Syracuse afterward, shouting "Eureka!" The citizens of Syracuse must have been agog at this nude Einstein in the street—why else would we know the story? Archimedes himself forgot he was naked, of course. If one doesn't have a body, one can't be naked in it.

The bathing of Archimedes was emblematic of Hiero's relations in general with his brainy cousin. A farsighted governor, he recognized the brilliance of Archimedes's theories, but throughout his peaceful reign encouraged the geometer to be more practical, to make his discoveries useful, as Plutarch relates, "by applying them through the medium of the senses to the needs of everyday life." Thus enlisted in the cause of reality, Archimedes adapted to ordinary purposes the pulley, the lever, and the screw. And so when Hiero took note of the rising power of Rome, he induced Archimedes to apply his skill in the realm of weaponry, to perfect the catapult and systematize its use for the defense of the city. Thus was Archimedes borne down into the world of weapons as he had been into his bath.

Even the brilliance of Archimedes couldn't save Syracuse in the end. Hiero died, after a half-century of festivals and public ceremonies. By then Syracuse itself had fallen into chaos. Hiero's fifteen-year-old grandson Hieronymus assumed power briefly, and was assassinated. The Romans were quick to seize upon an ally's weakness. Soon the Roman fleet was in the harbor, under the potent command of the consul Marcellus, the troops crowding the decks and bristling with spears.

The Romans weren't prepared for the shock they were about to receive. The power of the world, at that time, was still attributed to magical forces, even in Rome. Math at the time was a fine art, a parlor game, and the Romans still placed great stock in auguries—the reading of the appearances and flights of birds as indications of the intent of the gods. Roman consuls who defied the priests and their auguries did so at the peril of their positions in the government. So when the Roman fleet under Marcellus bore in upon Syracuse for the kill, it must have seemed that they had angered the gods somehow. Approaching the walls of the citadel, they were struck down, their ships broken and sunk, by huge stones howling out of a clear sky. Writes Plutarch, the Romans "began to believe that they were fighting against a supernatural enemy, as they found themselves constantly struck down by opponents whom they could never see." The stones seemed to come directly from heaven, flung by some huge hand, and the roar of the falling rocks alone was enough to unstring their wits. Behind the walls, out of sight of the legions, the old man Archimedes himself directed the silent firing of the engines he had devised.

He'd deployed hundreds of catapults, varying in size and function. Boulders showered down on the invaders, splintering their decks and crushing crowds of troops at a single blow. The Romans beat an initial retreat, regrouped, and attempted a counterattack that night, sneaking inside the set range of the big catapults, where they found that Archimedes had also designed a set of short-trajectory engines, which pelted smaller boulders and heavy darts onto the Romans as they came near the walls. For any ship that managed to withstand this heavy rain and close in upon the city, Archimedes had devised enormous cranes that, when run out from the walls on thick beams, gripped entire ships in their claws, hauled them into the air, and dashed them against the rocks. By this time the invading troops were so unnerved by these magical engines of Archimedes that the appearance of a single plank or rope on the walls sent them for cover, yelling in fright.

So Marcellus gave up, for the moment. According to Plutarch, he retreated out of range, ridiculed his own engineers and compared Archimedes to Briareus, the hundred-handed giant of Greek myth. Then the consul began a two-year conquest of Sicily, and in the meantime blockaded the town. The Romans did not hurry, as a matter of strategy—even Caesar's *Veni, vidi, vici* implies patience and observation in its middle term.

When at last he had the city surrounded, Marcellus resorted to bribery. He paid off a mercenary fighting for the Syracusans—promising him land in the conquered province—and the man left an entrance unbolted and a tower unguarded during a feast day. Once inside, the Roman swordsmen spread through the city killing people, among them—though Marcellus had ordered he be taken alive—Archimedes himself.

Two stories persist describing the geometer's death. In one, invading soldiers catch him fleeing from his house with a box of scientific equipment—a device for measuring the distance to the sun, among other things. The soldiers notice how tightly Archimedes holds the box, and assume it contains gold. Greedy, they kill him for his possessions, to find in the box only objects they cannot understand.

In the other story, which I choose to believe, a soldier sent by Marcellus finds Archimedes in his study, working on an abstract problem. Archimedes won't leave his desk until he delivers one last theory into the world by working out its proof. The soldier, a man of action infuriated by the mathematician's refusal to face facts, at last draws his sword and slays this abstracted creature, severing his last theory from the realm of the real.

The Romans took the city, and held it for the duration of their empire. Among other changes, they tried to adapt the Syracusans' Greek amphitheater to their own particular taste in drama—blood sport. They cut antechambers into the stone at stage right and left, out of which the opposing gladiators might emerge. But the Greek theater, built to represent the gods in tragedy, to awe and not to entertain, was never satisfactory for Roman purposes. The Greek stage, for one thing, commanded the crowd too fully. Ultimately the Romans built their own circular arena nearby, where they might sit godlike above the fray and judge the slaughter. As for Syracuse, once fallen, it would be provincial ever after, never again a world power.

Looking for the beginning of the catapult in history, I found it attributed to Syracuse most authoritatively, but discovered it in other places as well, and I began wondering if I hadn't gotten the terms wrong: not that history produced the catapult, but rather that the catapult produced history. Catapults made empires that kept records, and defended those empires so that it was more likely that such records might endure for posterity. Perhaps, too, later writers simply credited their ancestors with inventing the catapult to confer the prestige of the technology upon their own cultures. In any case, old catapult stories appear in various places, usually with the historical implausibility and psychological cogency of myth.

The Bible records the earliest catapult, at 2 *Chronicles* 26:15, a chapter describing the glory and might of the reign of Uzziah over the kingdom of Judah: "And in Jerusalem, he made engines of war invented by cunning men, to be on the towers and on the corners, for the purpose of shooting arrows and great stones." Uzziah lived, reigned, and died about 2,800 years ago, 400 years before the rise of Syracuse. The ancient Hebrew king himself came to a bad end. Perhaps in part because he was fortified by such engines, he began believing in his own glory, and entered the temple to burn his incense to the Lord in person. The priests were outraged. No profane man, even the king, could make offerings in the temple. They ordered him out of the sanctuary, but Uzziah the Mighty refused—he stood there with his censer in his hand and defied them. He needed no caste of priests to embody him, to represent him before God. But God, as it turned out, would have none of it. As Uzziah stood defiantly beside the altar, the Lord smote him with leprosy on the forehead. Thus struck down by God's bolt from the blue, the horrified king ran out of the temple, remaining a leper until the day of his death, after which he suffered the fate of tyrants when he had his story recorded by rivals who outlived him.

Whenever the catapult originated, once Alexander the Great first employed great offensive engines, the weapon's range and invisible potency made modern empire possible and transformed Western culture. Before such catapults, a fortified city was relatively safe. The Achaeans in Homer's poem tested the walls of Troy for ten years. But the strength of the old walls and the individual combatants was not on the same scale as the power of the new weapon. Flavius Josephus, in his account of the siege of Jerusalem in the first century A.D., writes that the Roman siege engines were so powerful that they could be fired far from the walls, where no one on the battlements could see them, "and so what was thrown by them was hard to be avoided." One man could not match this power, nor could several. "No body of men could be so strong," writes Josephus, "as not to be overthrown to the last rank by the largeness of the stones."

In this sense the invention of the catapult—and this shock at being so overmatched—was a psychological watershed for humanity, and a difficult one for us to imagine, living in a world animated by the unseen power of machines. With the great catapult, the new power of the engine was turned upon human beings, impressing them with engineering as never before. Not that human beings hadn't always been overmatched, of course; but until that point such huge and invisible forces seemed divine.

The catapult arrived with new values attached, diminishing the importance of localities and individuals. Witnessing the catapults that had been arrayed against him, one commander cried out, "Oh Hercules! Human martial valor is of no use anymore." The fighting spirit of the soldiers, like that of the Tommies who dashed out of the trenches of Flanders, shouting oaths to King and Country, suddenly became something more like an idle formality. Relatively speaking, human beings were exponentially weaker, both physically and socially, in the presence of these engines of war, and what strength they had began to be valued to the degree to which they had control over the machine. It was part of the process by which, as Gibbon writes, war "was gradually improved into an art, and degraded into a trade."

With the onset of specialized military engines, as Soedel and Foley note, "the equality of arms was lost." The new need for military engineers, superior to ordinary warriors no matter how physically able they might be, reinforced the social hierarchy in the military. And rigorous hierarchy in the military tended to spread into the imperial society as a whole. Thus the engineer not only invented technology—as a kind of unconscious by-product he encouraged a new relation of men to things and to one another. "In truth," writes Plutarch of the battle for Syracuse, "all the rest of the Syracusans were no more than the body in the batteries of Archimedes, and he was the directing and controlling mind." So Archimedes had his body, after all, one with many legs and arms and many heads.

One may draw a line at the great catapults, to distinguish an old world from a new one. After its invention, the original technology of the catapult flowered and spread from culture to culture, often via the medium of conquest. There were many local versions, and a foliation of names for these variations: ballista, beugle, blida, bricole, calabra, *engin à verge*, espringale, fronda, fundibulum, manganum, martinet, matafunda, petrary, robinet, scorpion, and tormentum, among others.

But at the close of the Roman empire, the technology of the catapult, like much classical knowledge, was scrambled or lost in the Dark Ages, and the powerful and exact Roman machines were replaced by cruder forms, the constructions of the people in a fallen time. It was not until the twelfth century that northern Europeans made widespread use of a large siege engine, the clumsy trebuchet, with its long arm and its heavy counterweight—usually a box of stones. Imprecise, immovable, liable to break down and generally stupid, the trebuchet was nonetheless formidable, powerful enough to throw a dead horse over a wall—as a kind of primitive chemical warfare—or, for that matter, to return the head of the hapless emissary of the besieged to his people.

In the fourteenth century came the cannon, and the power of weaponry took another exponential step away from the scale of individual human strength. One last catapult was built in America by the invading forces of Cortés in Mexico in 1521, when the conquistadors found themselves out of cannonballs in the New World. The machine destroyed itself on the initial shot, it turned out, by launching its first boulder too steeply. Later, in the eighteenth century, a British general and antiquarian would team up to build another anachronistic catapult, this one employed at Gibraltar to throw rocks at some Spaniards ensconced on a ledge below them which was for some reason unreachable by mortar fire.

The engines of Archimedes failed Syracuse—and failed Archimedes himself, for that matter. After his death, Archimedes gained fame throughout the Empire for his wonders of geometry. Rome adapted his catapults. And it was a good thing, too, that his name was honored abroad, because it might have been forgotten at home. A man of abstraction even in death, he had requested that his tomb bear only the diagram of a cylinder enclosing a sphere, with his formula describing the ratio of their volumes. A couple of centuries later, Cicero discovered this tomb in ruins—weedy, the diagram worn off, the shrine neglected by the conquered citizens of Syracuse.

CHAPTER 7

The Springs

In West Oakland, Harry wanted me to help rip the carpet out of his new living room, and I was not bringing up the catapult, though my silence grew more anxious by the day. I waited as long as I could, and even so, Harry complained when I finally came out with it. "We've got to stop goofing off," I said. "We've got a deadline." It wasn't a good thing to say. It made him mad.

"You call this goofing off?" he said. "We've got all kinds of stuff still in boxes—I'm living with boxes in my living room—so don't give me deadline." He huffed. "We haven't even moved in here yet."

"You've moved," I insisted. Everything he had was in the new house, wasn't it?

That wasn't the point, Harry said. The place still felt like somebody else's house. That rug, he said, pointing to the tangerine wall-to-wall carpet in the living room, that rug had to go.

"I've heard all I want to hear about that rug," I said.

"Look," said Harry. "There's a difference between just moving and moving in."

"And just how long is moving in going to take?" I said. Harry didn't know. He just knew he had to get the rug up.

"I want to see the wood under there," he said.

On my way home I began to wonder how I might proceed without Harry. Before his house had intervened, he'd told me that our first step would be to buy some springs. At home that night, I tried to read up on springs at least, poring over the closest thing to a mechanical reference book in my house, the *Oxford English Dictionary*. The word *spring*, I learned, was an ancient one, carried forth from the Germanic tribal past of the English language. Over the ages, it had flourished into dozens of meanings that now spread over eight pages in the big book. I scanned them with the magnifying glass that came with the set. By the Dark Ages, *spring* was already both noun and verb, literal and figurative. It occurred in Beowulf and Chaucer and Shakespeare, and originally indicated a water source, that place where something seemed to spring from nothing. Later it came to describe the dawn, the first season of the year, and the appearance of new shoots or sprouts. Its meaning as the name of a mechanical contrivance probably came by association with young trees—sprigs, the first springs to be employed as engines. It occurred to me that a spring, used as a bow or a snare for instance, stood ahead of the wheel, and just behind a thrown rock as primal technology. From the energy stored in the resilience of wood came the first springs, from which all guns followed, not to mention all violins.

And I remembered a copse of cypresses in the Headlands, planted against the nearly constant northwesterlies off the ocean. The force of the wind there rivals gravity itself as a steady demand on the trees, and they have grown permanently flung out to leeward, the whole hillside of them holding that gesture, like the memory of the wind, on calm days. Springs are made like this, I thought, by some original resistance to the constant powers of the earth. The tree grows against gravity and bears up against the prevailing breeze. The wood in the bow possessed that old strength that the tree had summoned against the wind. Springs, I thought, had the power of habit.

In the morning I felt ready to rely on my instincts and act.

I'd gotten the money, hadn't I? By then I'd put the $500 into my account. Nothing was stopping me from going out and buying some springs for the catapult, I thought, though I wasn't going back into that dark and grimy spring shop on East 14th. I didn't want to face that ancient character alone. So I found other spring shops in the phone book, among them the Ashby Spring Manufacturing Company. The place was in Emeryville, near Harry's neighborhood. I got in the car and drove over there.

Compared to the ancient character's place, the Ashby Spring shop was like a hangar, a broad pad of concrete beneath a high girdered ceiling. The two guys in the shop wore identical brown coveralls and just nodded at me when I told them that I wanted to look at some springs.

"Are you an artist?" one guy asked. I shrugged. The guys in brown seemed to know about artists. Did I know Abduljaami? the other asked. I did. Down the street from the spring shop was Abduljaami's Wooden Wonderland, a bizarre and crowded menagerie sculpted out of huge blocks of walnut and painted in wild colors. Abduljaami was a big bearded black man who raised chickens beneath the freeway and cut his sculpture in the front yard with an axe and a chainsaw. These two guys in their brown coveralls said they played poker with Abduljaami down at the Oaks Club.

Idly, I said that a lot of the warehouses around there seemed to have been converted to studios. "We know," said the first guy. "They take stuff out of our dumpsters."

So I decided not to tell them what I wanted the springs for. Still, as I chatted with them, I felt pleased to be proceeding without Harry. Who needed him, anyway? These Ashby Spring guys gave me a friendly tour of their shop. They evidently spent their day coiling various thicknesses of wire into springs, then baking the coils to temper the steel. One guy showed me big black springs that looked like heavy bracelets. They were dense, the wire tightly coiled ten or twelve times and weighing about as much as a camera. Plus they looked good—they had been sprayed with black enamel, and they were shiny.

The guys in brown were making a huge batch of these coils on a Defense Department contract for the air force, I was delighted to learn. The irony of it, I thought.

"They retract the arm on a pig," the first guy said. A pig turned out to be one of those squat tractors that tow planes around on runways. I felt like I'd come to the right place.

"Nobody else in the country makes these particular springs," said the other guy. They were in the process of making a thousand gross of them, he said—144,000 springs. It was an impressive number of anything—it made me remember a medieval poem in which a procession of 144,000 virgins appear, called the Host of the Innocent. But I didn't see that many springs in the shop. Maybe they shipped them out as soon as they made them.

"You know," said the first guy, "you may just be in luck."

We went to the back of the shop, where they pulled a big cardboard box from underneath the workbench. It was full of black bracelet springs. He put a pair of them in my hands. "Just by chance," said the guy, "we happen to have made a few extra of these."

"We're not supposed to sell them," said the other guy.

"Come on," said the first one. "We can make this guy a deal, can't we?"

I bought them on the spot—four of the black, shiny bracelets for sixty dollars. I wrote them a check, and one of the guys in brown even insisted upon putting them in a cardboard box, and carrying them out to the car for me. Harry's shop was only a block away, and I went right over there with my prize purchases.

I found Harry working with his spray gun, painting a set of shelves that he'd laid out on oil drums beneath two huge roaring fans that pulled the fumes out the windows. He had on what looked like a gas mask. Inside the shop every surface—

the walls, the floor, the ceiling, the windows, and Harry him-self—had been sprayed a uniform gray. I yelled to him over the howl of the fans and the air compressor, and proudly held up my springs. Harry looked up at me and pulled off the gray snout of the mask. Slowly he cut the equipment off, switch by switch. "What do you call those?" he said finally. The paint around his eyes made him look like a raccoon.

"Our catapult springs, Harry," I said.

"Those won't work," he said. First of all, he said, they were engineered for something we weren't going to use them for—we couldn't attach them to anything. Plus they weren't even uniform, he said—they were probably seconds, cast-offs, re-jects. And most of all, they were tiny. I just didn't understand how much power we were going to need, Harry said. We had big rocks and we were going to need great big springs, capable of generating thousands of pounds of force. These springs were puny. When I looked at them again, they did look puny.

Worst of all, Harry guffawed in my face when I told him I'd paid sixty dollars for the springs. Some of the gleam went off my successful grant-writing at that moment. The bracelet springs were worth at most about $2.50 apiece, Harry said. I told him about the guys in brown. Harry said I'd probably made their day.

And when I got back to Ashby Spring, I saw that he was right. The guys in brown were headed out the door in a fine mood, going to lunch early. They seemed to have spent the past hour dancing a jig. They were probably going down to the Oaks Club with my sixty dollars, I thought, ready to parlay their morning windfall at the card tables. They stopped when they saw me, and one of them growled a little, but they didn't argue when I asked them for my money back, handing them the box of shiny springs. The whole thing had been too good to be true, they seemed to be thinking. They gave me my check and told me to come back anytime. Not without Harry, I thought.

* * *

Harry must have had the same thought, because by the next weekend he was ready to do something about the catapult. Saturday morning I went with him down the peninsula to pick up his son Isaac, a thoughtful, red-haired kid who lives with his mother in Burlingame during the week. Isaac was often quiet when we picked him up, existing somewhere between saying good-bye to his mother and hello to his father. But that day we cheered him for a moment by announcing that we were stopping off at a cool place on the way home. Where? he asked. "A steelyard," said Harry.

"Oh," said Isaac.

We drove up 101, a funky bayside strip of freeway, and pulled off into an area of salt flats and industrial buildings along the shoreline. One of these was a steel wholesaler and scrapyard that Harry knew about. In the yard sprawled stacks of steel, raw plates and beams and formed steel housings, transformers, drums, boilers, and other odd stuff: a jet fighter's ejection seat, an autoclave, a diving bell.

Isaac and I moved warily through the stacks, but Harry hurried ahead, moving eagerly from thing to thing, pounding and slapping and rubbing them. "Whoa!" he shouted. "Look at this weird stuff in here! Just look at the electricity that goes through that thing!" He acted as if he were hungry for the ferrous scrap. "Look at these casters!" he said. "Check this out—solid ingots!"

Isaac and I followed Harry around. To us it was just stuff. Stuff and more stuff, ubiquitous hulking steel stuff, that does something but you don't care what. Sharp, heavy stuff. "Come on, Harry," I complained. "Let's look for the springs."

"Don't touch everything, Papa," Isaac said.

But Harry was excited. He headed deeper into the scrapyard, climbing on top of the piles, pushing things around to get a better look, disappearing down the canyons between the stacks. We could hear him in there, exclaiming and explaining, "God—look at this thing! Look at these tubes! Look at the rust on this thing! You know how much this thing weighs? Look at this—how many times in your life do you get to see things like this?" Nothing and everything in that yard looked like a catapult to me.

When Harry's enthusiasm for the scrap finally wound down, we went inside, where the place was like a lumberyard for steel, just racks of unmarked beams and rods and ingots and plates. We asked about springs. The salesman told us that they supplied steel to people who made springs by tempering them. We'd have to find a spring guy, he said.

So we made our way back out of the steelyard. Harry said he'd known about the place since he was a kid. Harry had gone to high school down there on the peninsula, running around in the woods before the freeways went in, and roaming and scavenging these flats by the bay. On our way out of the steelyard, he looked through the fence out at the salt flats and remembered that he had once shot an abandoned television with an arrow out there. The first two shots had just bounced off, he said, but when he finally pierced the picture tube, the thing had imploded. He made the sound of the tube imploding with his mouth. Isaac listened to his father and gazed out through the fence across the littered marsh. "Do you think it's still out there?" he said. "Sure," said Harry.

The steelyard adventure had left Harry in a fine, talkative mood. On the way home in the car, he talked about steel. He regaled us with a description of the process of oxidization. "Rust is another form of fire, you know," he said.

And as I got out of the car, Isaac gave me a warning. "You better watch out, Jim," he said. "Papa will just take over."

The next weekend we again went spring-hunting, this time taking along Susan's son, Ross, a towheaded and frenetic kid. At the steelyard we'd found that it would be too difficult to make our springs out of raw steel, so this time we went to an auto junkyard, looking for some kind of ready-made springs that might suit our purpose. Maybe car suspension, Harry thought.

Ross is usually in motion—he is a natural athlete already—but even though he loved the junkyard, he moved carefully there. We walked through the yard between the rows of junked cars stacked three high, some of them horribly wrecked, their body metal shredded. All of the cars roofs had been squashed, to make them easier to stack. The clerk, styled as a punk in black, walked us through the yard, and he and Harry pointed out various things amid the wreckage. There was a car with its steering assemblage exposed, so we could see what happened to the mechanism behind the dashboard. We found a transmission housing, like an oversized French horn case, a gearshift lever attached to its pointy snout. I was never a car mechanic, and it seemed a little surreal to see something familiar connected to its usually invisible working parts, which were larger than I had imagined and somehow gruesome.

The clerk took us into a building at the back of the yard where piles of cannibalized car parts lay on tall shelves and in heaps on the dim warehouse floor. Ross took my hand. We worked our way back into the spring section, and the punk clerk gestured to a pile that looked like part of an ancient armory, a clutter of bows in the dark, as if in some nearby hall warriors were singing and drinking mead under torchlight.

These were leaf springs, resilient steel leaves stacked and fixed. I picked up one of the grimy bows, twenty pounds of steel. "That's a good one," said the clerk. "That's from an RX-7."

I wanted to buy these car springs right away. But Harry looked them over and put them back in the pile and we left. Those were closer, he said in the car. Leaf springs would work. We wouldn't have to use torsion springs. Until then, I hadn't quite understood what Harry had meant when he talked about leaf springs and torsion springs. To me, springs were coils, like bedsprings. But leaf springs were flexible bows. And torsion springs, like the bundles of sinews on the Roman catapult, released power by untwisting. Torsion springs would have been tricky, said Harry.

Plus, he said, we could make our own leaf springs from individual leaves. We couldn't use raw steel—What are we going to do? said Harry. Stick it in the oven?—and the car springs had been wrong because they were already too adapted. There wasn't any easy way to remake them into what we needed. What we needed were springs in some intermediate state—not raw steel, not assembled product. And Harry wanted big springs—for trucks, not cars—to yield the power we needed.

"Where could we get those?" I said.

"My brother-in-law," said Harry. "He's in the business."

CHAPTER 8

The Torch

The next night after work Harry drove over in his battered white Ford pickup, a miserable vehicle that had been smashed at various points around the body. The rear bumper had been bent under the bed of the truck; the driver's door had been smashed in; the hood was crumpled. Once somebody crunched the door, and for weeks afterward Harry had simply climbed in through the other door over the gearshift lever. The situation finally drove him into a fury, and, refused access one last time, he pummeled the damaged door with his fists until it came open.

The truck's internal parts were almost as bad as its looks. The ignition cables were quite loose, nearly severed, and Harry had to reach through the hole where the radio should have been to hold the wires in contact as he turned the key. Harry had rarely, if ever, changed the oil in the truck. When he burned up his first engine, he just had a new one put in. A truck is a truck, says Harry, a pile of metal on wheels. I'd always been more personal with my car, as if it could feel my beneficence. I'm meticulous about its grooming, occasionally cleaning the dashboard with a toothbrush. The floor of Harry's truck was ankle deep in old rope and padding, invoices and cans. But, oddly enough, I always felt relieved to go anywhere in Harry's truck. To get into the littered and dingy cab was somehow to live for the moment in disdain of sheer stuff, to know that you could do no harm by spilling more paint on the seat or kicking the dashboard. It was as if you could roll the truck's stinking hulk over a cliff somewhere and just walk away, dusting off your hands.

Talkative as usual, Harry drove through the Mission up Van Ness Avenue to Bachman's Welding, where his brother-in-law David worked. David did have truck springs, he informed me—matter-of-factly, as if announcing that so-and-so had a cup of sugar we could borrow. And Harry said that David was going to let us use his acetylene torch as well. A torch I didn't know about. But Harry was excited at the prospect.

"Are we going to cut the springs with a torch?" I said, a little nervous.

"Can't cut springs with a torch," Harry said. "It rearranges the molecules and they lose their temper. Once you heat them, springs go blah in that place."

"Great," I said. "So what are we doing with the torch, then?"

"We need to cut some I-beam for the mounts," Harry said, adding, "Never used a torch, eh?" He laughed, a squawk. "Don't worry, I'll do the cutting."

Harry reminisced a little about his experiences with the torch. He'd learned to weld in his father's shop, welded in art school, and once put up scaffolding for some welders inside the hull of an oil tanker. He was never so petrified, he said. When I couldn't take Harry's welding stories anymore, I finally insisted that I was going to use the torch, too.

"Whatever," said Harry. "Could be dangerous, though."

"I'm doing it," I said.

"Could be deadly," said Harry, squawking again.

Bachman's was a holdout on that pricey strip, just off Van Ness at Pine. It shared the avenue with multiplex cinemas,

BMW dealerships, and the Hard Rock Café with its line of children out front. Harry said that Bachman's was more than fifty years old, that Old Bachman had built the place like a battleship, all by himself. He'd fabricated all his own tools, even, Harry said. But in fact Bachman's was doomed at this point. David had told Harry that the building was in the process of being sold to Japanese investors—for something like two million dollars. "Just think of it," I said. "No more welding on Van Ness Avenue. Big deal."

It was getting dark as we pulled up; David's black Alfa was parked out front. The welding shop, an old orange storefront, was overhung with painted signs and a sheet-metal robot holding a welding torch. The signs over the door of the place had a circus-poster quality—AXLES STRAIGHTENED and HOT OR COLD METHOD—and made the establishment seem like a blacksmith's shop for cars. Beneath the signs stood an orange steel door big enough to drive a truck through, with another person-sized orange door within it. Harry tried them both, found them locked, then banged on the big door, first with his open palm, and then with one of our catapult stones, which he'd brought along to help us determine the proportions of the springs, to provide scale. When the boom of the steel quieted down, we heard David inside mumbling something. The small door opened and we went in.

David was lean and silent, a chain-smoker. He had a cigarette in his hand when he opened the door, and gestured with it, saying nothing, but leading us back through the long chamber of the shop's ground floor. Behind the dark workbenches and crannies of tools and tall canisters, racks of steel covered the walls, and almost everything under the fluorescent lights in the shop was the same color, oily gray. Amidst the gray clutter, two gold Mercedes sedans stood propped on hydraulic lifts for the night, their door handles elevated to eye level. David took us into the office, a booth really, jammed with catalogues for parts.

I asked David how long he'd worked there, and he said he'd

begun working at the shop as a welder, but that lately he was more of a front man for the place, greeting the customers, listening to them talk about what they needed, giving them estimates. Actually, he said, he was doing more design now than anything else, working with Sandra, his wife and Harry's sister. Harry had made a rather elaborate drawing of our catapult-to-be, but he'd forgotten to bring it, and had to sketch another quickly on a pad at David's desk. His sketch showed what looked like a winged weapon, the only worked detail the intersection of the bow and the stock. Harry described to David the other steel pieces he wanted as mounts for the springs, and handed him the rock he had brought along.

"You forgot your drawing, but you remembered this?" David said.

"It's all we need, really," said Harry.

As far as I was concerned, it was a good-sized chunk of granite, an impressive rock. But David just clucked a little and looked at the rock like it was a trivial thing to work steel over. He didn't say anything, but I felt he was humoring us, perhaps for the sake of his family relationships. He just told us that we'd need some really big springs, and added that the whole business could be risky.

"You have springs that big?" I said.

"Oh, yes," he said.

David took us back out onto the shop floor, where the gold cars stood on their lifts. The shop did a lot of custom car work, he told us, specializing in beefing up steering and suspension systems to absorb the extra shock of driving on hills in San Francisco. Both the Mercedeses had come from a single customer, David said, who had taken them directly from the dealer to the shop. On their stands the cars looked like gold toys.

We followed David into a big freight elevator. He slammed the steel gate and we descended into the basement. "This thing hydraulic?" Harry said, looking at the empty concrete shaft overhead.

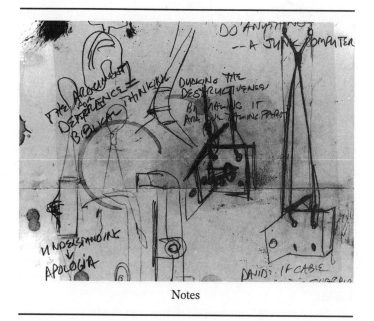

Notes

"Yeah," said David. "There's a huge piston under the floor. This thing will lift anything," he said.

In the basement, we stood a minute in dense darkness and the smell of oil as David found the light switch, then illuminated a cavernous space stacked with steel, different kinds of beams and pipes in nests and piles. We made our way back through the stuff, past open barrels of black oil, and when we got near the back, David gestured to a space along the wall. In the alcove, nestled in stacks like spoons, lay several dozen bowed steel blades, individual leaves for truck springs, each thick as toast and about four feet long. The mother lode, I thought.

Harry waded right in. He pried his fingers under the topmost blade in one of the stacks and came up with it, retrieving it in both hands as if it were a big fish. He took it to a clear space on the concrete and pushed one end of the long leaf into the floor, trying to make the steel give. He strained into the piece with both arms and it barely quivered. I got a leaf of my own from the pile. It was heavy enough to be unwieldy, and when I put my weight onto it, it felt as unbending as a post. "Believe those might do," David said. Harry and I each took two of these heavy leaves, slung them across our shoulders, and bore them back into the elevator like dwarves with booty.

Upstairs, we lay them on the floor with the stone we'd brought. "Look at that," said Harry in satisfaction. The proportions seemed right, like an arm's to a baseball. We were making an effigy of a human being, it seemed, one stripped of all capacities except stone-throwing, and that capacity amplified, as if in compensation for whatever else a person was. And something else impressed me in this configuration of bow and stone—that it was going to take tremendous force to cock the springs. "We're going to need some kind of huge winch," I said.

"And some big bolts to hold the springs down," said Harry, adding, "This is our engine—we're going to have to design around it.

"First things first, though," he said. Harry wanted to shorten two of the leaves, and David said that we could use the suicide saw, which was what he called Bachman's homemade chop saw, a spinning vertical blade, not really a saw so much as a thin grinding wheel on a hinged arm that chopped down into steel held in a vise beneath it. Harry could do all the chop-sawing he wanted, I said, and he was only too glad to do so. He measured and marked a leaf, fixed it in the vise, snapped the on/off switch, and lowered the blade. The leaf screamed and flung a furious fan of sparks over Harry's feet and into the wall behind the saw. Unfazed by the fury, Harry chop-sawed away, cutting the steel where he had marked it. David spread-eagled himself against the door panel of the nearest Mercedes, protecting its gold finish from the stray sparks. I stood back with my fingers in my ears.

When Harry had cut down two of the leaves, David gave us two steel bands, and we bound the leaves in pairs, a longer leaf against a shorter one that would support and strengthen it. Then we had two sets of finished springs, each consisting of three parts. Our first assemblies, I thought. To me, they had a rightness, like simple sentences.

Then it was time for the torch. David took us across the shop and gestured to a stack of I-beams on the floor. Compared to the springs these I-beams were thick and dull. I was more comfortable with this bovine steel. These guys hold up freeway overpasses, I thought. The springs might bite us, but these we could tame. Besides, I liked the simplicity of their name, and the letter they spelled in cross-section.

We grabbed a likely candidate—a beam that looked like a short piece of railroad track, and dragged it out of the pile onto the open floor where we could work on it. Harry measured and marked two sections off with a piece of copper rod. Then David went to the back of the shop and came back wearing heavy gloves and towing a primitive-looking iron-wheeled cart. On his head was a welding mask with its visor up, and he was carrying two others, which he gave to us. The cart carried two tanks, one red and one green, each sprouting a hose the same color as the tank, which came together in a single brass nozzle, and David held this nozzle up and began to narrate his actions tersely, like a fighter pilot.

"Green for oxygen, red for acetylene," he pointed out. "Check gauges." On the handle of the nozzle were three dials and an elongated trigger. He turned the dials. I was impressed by David's expertise. Harry just looked impatient.

"Primary oxygen at a hundred percent," David said. "Crack acetylene. Crack secondary oxygen." The nozzle began to make the hollow hissing of a small jet. David picked up a big spring-loaded clip. "Light torch," he said. He snapped a spark from the clip and ignited the gas. From the nozzle leapt a blue flame, about eight inches long and wickedly pointed. "Ready," said David, flapping down his faceplate. "Don't look at the torch," he shouted behind his mask. Harry had told me about this—the brightness, he'd said, could burn your retinas. But I couldn't not look.

Harry and I put on our masks. These were basically headbands with hinges at the front, on which the heavy, nearly featureless face masks could be raised and lowered. I lowered mine, and peered out at the scene through the dark glass of the mask's little window. David advanced on the beam, which lay propped on some scrap steel at our feet. He knelt over it and shouted at us behind his mask. "Squeeze the trigger. Hold the flame a couple of inches from the steel." He did this. The jet grew louder. The flame poured over the beam like a stream of liquid, at first to no effect, then igniting a fountain of sparks, and then finally ripping open a slot through the steel. The molten steel spattered and fumed bitterly, reeking like burnt hair.

The violence was impressive. Steel clearly doesn't want to conform to human purposes, I thought. Wood you can work with your body's own strength, but steel needs deadly forces to come into being, flame so hot it took millions of years for us to manage it. Even through my mask I could feel that heat. You could die doing this, I realized.

Once David got the cut going, he released the trigger, extinguishing the shower of sparks, and stood up, the torch still hissing hollowly, its blue blade steady. "You guys do this," he said. "I do this all day." Harry wanted the torch, and I was glad to let him have it. Harry and David gingerly exchanged the gloves and passed the nozzle with its hot jet between them.

Harry knelt over the I-beam, his pale back exposed above the waistband of his jeans, and reignited the blast over David's cut. Sparks bounced off his mask and gloves, striking also his chest and arms. Beneath this fiery bombardment, tense and frantic Harry cut the thing, cursing through his mask when the cut wandered off the line we had scratched into the steel. When he finally cut through it, the end of the red-hot I-beam fell off, and Harry had to jump back to keep it away from him. In the process he stepped on a bit of molten slag which instantly melted into the sole of his sneaker, stinking. He shook the smoking thing off his foot, and lifted his mask. Behind it, his face was elated. "Your turn," he said, indicating me with the torch.

I took the gloves from him first and put them on. They were too big for me, and damp inside from the hands of the other men. When I took the torch from Harry, I took care to get a good grip through the leather, with both hands on the nozzle. I held it well away from my body, its sharp blue glow dim through the little window. At this odd remove from things— behind leather and steel and glass—and yet completely concentrated on them, I moved across the shop floor like a deep-sea diver and knelt over the beam.

The flame, when I squeezed the trigger on the nozzle, jumped from its compact blue form to a long flickering one, very bright even behind the glass. I put that flickering tongue to the steel. As the beam heated up, sparks shot into my arms and hair, and beads of molten steel began to form on the surface of the beam. A cut began to open at its thick lip. Harry leaned over me, shouting, "Stay on the line" through his mask. But it wasn't easy to control the flame, which was pouring through the steel the way hot tap water pours through a chunk of ice.

Harry wanted the torch back, I could feel it. He hovered there above me, wanting me to give up and let him finish the cut. This more than anything else made me bear down and finish. The sparks sizzled the hair on my arms; my gloves seemed immersed in the fire. My flame wavered around the line we had drawn on the steel. I had two inches to go, then one inch, then the weight of the unsupported end of the beam began to spread the glowing cut. And I finally hacked off the last bit, and danced away from the hot end when it fell. Then

I gave the torch to David and he shut it off and for a while there I was exultant.

We set up the I-beam sections—still hot—with the spring assemblies, and posed the whole thing with the rock again, getting our first look at the scale of our weapon-to-be. Beside that steel, the stone had shrunk, and what I had thought impressive, too heavy to throw, seemed laughable, golfball-sized almost, no match for the mechanism. We heaped scorn upon the rock, our enemy, our puny victim.

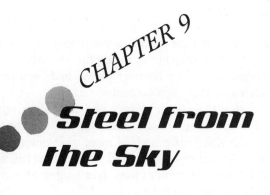

CHAPTER 9

Steel from the Sky

In 1803, on a clear April night, a particularly dense and fiery display of meteorites disrupted the steady pattern of the stars. With a sound "like pistols or distant guns at sea," they burst through the atmosphere, and were found in quantity and examined by gentlemen of the Enlightenment. Two gentlemen from Massachusetts prompted wide outrage by claiming that the objects were "stones from the sky." Thomas Jefferson himself wrote a refutation. It was "easier to believe that Yankee Professors would lie," he said, "than that stones would fall from heaven." Nonetheless they were rocks from the sky, and given such a large number of them, the scientists of the day were able to draw their first conclusions about the composition of some falling stars. They found celestial iron.

Perfectly smelted in the core of some shattered planet, this iron was crystalline in structure, and contained a high proportion of nickel. At that time its high nickel content aroused interest mostly among scholars of prehistoric humanity, who for some time had been wrestling with the existence of very ancient iron tools. They had already concluded, rightly, that iron in its unoxidized metallic state does not exist naturally on earth. So the men inventing the smelter seemed to have invented iron when they ushered in what was called the Iron

Age. But there were articles of iron dating back as far as 4000 B.C., twenty-five centuries before the Iron Age. Most of these remains, shaped by hammer blows, had been preserved for posterity by their high content of nickel.

So after the meteor shower of 1803, some scholars concluded that those early artifacts were in fact composed of iron from the sky. This prompted more wide outrage, until some years later anthropologists would find that the Eskimos, among other Native American tribes, had long used extraterrestrial iron blades on their bone tools and spears. One tribe in Greenland had chipped their iron from a single huge meteorite—the people made an annual pilgrimage across the ice to reach this metallic stone, which they called "The Mother." And later scholars of the English language would posit that the root of the word *iron* was related to the root of the word *heaven*.

But it remained for humanity to find a way to produce this celestial substance on earth. Early iron was rare and precious, so much so that it was sometimes beaten into ingots and used as currency. Humanity's functional relationship to the metal began with the geographical diffusion of smelting technology, with this new ability to build and control fires of great heat, heat as great as that within the planetary cores that had formed the iron in meteorites. With such heat, people suddenly had vast supplies of potential metallic iron in the surface iron oxides of the earth. In certain ferrous regions, in the mountains between the Mediterranean and Black seas, for instance, large campfires must have been accidentally smelting iron forever, and it was probably in such a place that the fire-making improvements necessary for purposeful smelting arose.

To smelt iron from ore, you dug a deep hearth high up on the hillside. Then you stacked your fuel and ore carefully and let it burn for hours up there in the wind. When the fire cooled you could pick through the ashes for the lumps of pig iron, soft carboniferous clots full of the ashes and leavings called clinker and gangue. Then you'd hammer and fire the lumps until you had something. The hammering would start with a thud and end with a clang.

The Hittite people arose on the Anatolian plateau, along the upper Euphrates in the mountains of what is now Turkey, between the Mediterranean and Black seas. The Hittites became a world power about 1500 B.C. in part because of their iron-making technology. In about 1300 B.C. the king of the Hittites responded to a request for iron from Ramses II, pharoah of Egypt. "It is a bad time to make iron," he wrote to the Pharoah, as if iron-making were seasonal. It probably was, actually, depending on the heating potential of the wood used for fuel. To soften the blow of his refusal, the Hittite king sent Ramses a gift. "Behold!" he wrote. "Now I am sending thee an iron dagger blade." Perhaps the Hittite king hoped that the pharoah would simply place the blade safely among his other jewels and forget about it as a weapon.

But the powerful knowledge of iron-making procedures would not be contained, though it was occasionally outlawed. The iron-working coastal people called the Philistines forbade the use of iron by Hebrews, their technologically inferior vassals who occupied the inland hills of Judah. Still, the information spread away from the Anatolian plateau into Asia, Africa, and Europe. By 1000 B.C. that iron-working region already included Greece and the Middle East. And by 500 B.C. the iron-making area encompassed the known world of the West—from Spain to Afghanistan, Ethiopia to the Baltic Sea. Today that region encompasses the globe, which might be known as the Iron World.

Steel is iron hardened and purified through diffusion by other elements. Such solid diffusion defies common sense— solid things seem to be separate things. But a sandwich of titanium and lead, for instance, would become a single solid in twenty thousand years, after which the bolts holding the sandwich together could be taken off and thrown away. Copper clamped to gold will contaminate it in just a couple of decades as the molecules mingle, but even this tries the patience of most people.

Inhuman temperature, though, can stand in for inhuman time. Heat speeds solid diffusion, weakening the bonds between atoms. Under great heat, carbon diffuses through iron quickly, carrying off the oxides and strengthening and hardening it. The first forged tools were simply case-hardened—the carbon in the ash of the first forged implements diffused only into the surface of the iron as the tool or weapon was heated and hammered. Later, blast furnaces with greater heat and closer control of the materials could diffuse carbon evenly throughout the iron. A kind of blast furnace was apparently used in India as early as 500 B.C., and produced the first man-made steel, which was in turn made into swords at Damascus. Wrought-iron swords would often have to be unbent between skirmishes on the battlefield. Swords from Damascus would sing, could cleave a piece of loose silk tossed in the air.

But it would be the fourteenth century in Europe before the forge came into general use, producing cannons and cathedral bells of bronze, among other things. Sometimes these furnaces even reworked the same metal into cannon or bell, depending on the state of war. Cast iron came later, but by the middle of the nineteenth century, many of the mechanisms and processes that would be called the Industrial Revolution were waiting on steel, which was still precious in modern terms. At that time, it was produced by a process called puddling in England, which required baking the smelted ore, pig iron, for a long time, until the carbon monoxide gas in the fuel diffused its carbon through the iron. By puddling, it took about two weeks to make fifty pounds of steel, about enough for a single piece of railroad track.

Puddling was made more expensive by its enormous use of fuel. England, the world's foremost iron-maker from the Middle Ages to the 1880s, cut down a good part of her forests for charcoal to feed the forges. As early as Shakespeare's time, the regent was issuing prohibitions against woodcutting for iron in Sussex and elsewhere. By the nineteenth century, the trees were gone, and the British were reducing their coal to coke for the furnaces, a process still wasteful and expensive.

In November of 1854, a middle-aged British inventor went to a French party with a mahogany bullet in his pocket. He was Henry Bessemer, a shrewd man with only a little knowledge of metallurgy. As a boy he had poured molten lead into type for his father, and he had made his first success in business as a teenager by figuring out a way to cast white metal reproductions of beetles and flowers.

The stunning news about spinning bullets—that they would go faster and pierce harder—had just burst upon the militarists of the period. Their guns were suddenly obsolete, smoothbores that couldn't spin anything. Here was his market niche, thought Bessemer. He proposed to groove the bullet instead of the barrel by cutting a kind of sluice into it, so that some of the exploding charge would be vented laterally, spinning the projectile just as it was fired.

He carved a model of his bullet out of mahogany, and took it to the house party in Paris, a Tolstoyan occasion of farewell to some French officers just assigned to the Crimea. Prince Napoleon, the cousin of Emperor Napoleon III, was in attendance, and Bessemer had himself introduced to the prince not as the Tin Beetle Man but as the inventor of an elongated projectile that could be spun out the barrel of a smoothbore cannon, of which the French had many. The prince was interested, and within a few days, Henry Bessemer was in audience with the emperor at the Tuileries. He came home with an advance on royalties.

And in only three weeks he was en route back to France with his sluiced bombshells. Overseen by a commandant in Vincennes, Bessemer fired his shells out of a French twelve-pounder cannon, shooting them point-blank through six thin wooden targets and into a snow-covered hill in the public park.

Bessemer had painted his projectiles with black varnish; scratched by the targets, the surface showed the marks of a spinning bullet. It was an impressive display, and Bessemer retired from the field confidently. But when Bessemer and his French hosts came in out of the cold and gathered by the fire in the low hearth, "under the happy influence of a steaming cup of good mulled claret," as Bessemer writes in his grandiose autobiography, the French commandant who had overseen the test-firing raised a single polite but devastating objection.

Bessemer's long projectiles were more than twice as heavy as the old cannonballs. Placed in the cast iron French cannons, they would add tremendous overpressures to the stress on the chamber. There was no getting around the implication: Anyone who put such a projectile into an ordinary, battle-worn French twelve-pounder would be a great deal more likely to blow himself to bits with his own gun.

This time it was a lonely journey home to England, but on the way Bessemer resolved to press on with his project. His task was now much larger—he had to find a way to improve the quality of cast iron for cannons. To his credit, he was not merely undaunted by his ignorance of the metallurgy of iron— he was encouraged by it, in the best sanguinary English manner. "I had nothing to unlearn," he writes.

And within a year he claimed to have come up with the answer, a simple and violent idea. Oxygen could control the carbon content of steel, and would bear off other impurities. Why not, thought Bessemer, just blow a strong blast of air through a bath of molten iron and oxidize it rapidly? Because, he might have added, you tend to burn your house down if you try it.

Certainly Bessemer underestimated the violence of his first experiment. He constructed a small furnace with holes in the bottom, through which he could blast air with the help of a twelve-horsepower steam engine. The oven chamber of this furnace was about four feet high, "sufficiently tall and capacious," he expected, to prevent anything but a few sparks and heated gases from escaping from the hole in the top of the furnace. But when he set it up in the garden of his place at St. Pancras in London, and blasted the air through the molten bath, he produced an astonishing fireball.

Bessemer and the lad he employed to actually work the contraption ran for their lives, from this "veritable volcano in a state of active eruption," as Bessemer described it later. A voluminous white flame shot from the top of the furnace, accompanied by a series of explosions within it. Molten metal splashed high into the air. Bessemer and the lad took cover, and the machine went on erupting, its on/off switch unreachable beneath the rain of fire. After a time, the fire died down, and the two ventured out. Bessemer instructed the incredulous lad to tap the contents of the furnace into the ingot mold. "Out rushed a limpid stream of incandescent malleable iron," writes Bessemer, "almost too brilliant for the eye to rest upon." What Bessemer called malleable iron was a purified product, carbon evenly diffused throughout. He'd made instant steel.

Never a man to meditate, Bessemer rushed his process into patent and print. His description of it, entitled "The Manufacture of Iron Without Fuel," was squeezed in as a last-minute addition to the program of that year's meeting of the British Association for the Advancement of Science, and published in its entirety in the London *Times* two days later. Within a month, Bessemer had taken in about £27,000 in fees and royalties. He had dropped a bomb on the steel industry of his day, but there was a hitch.

The bomb turned out to be a dud.

None of the licensees in the Bessemer system could duplicate its success. All that came out of their new converters was useless, brittle, unworkable metal, inferior to the worst puddled product. His clients were furious and cried fraud. Bessemer was denounced in the press. He was at best, said one business columnist, a "wild enthusiast," and his invention was "a brilliant meteor that flitted across the metallurgical horizon for a short space, only to die out in a train of sparks and then vanish into total darkness."

But Bessemer's struggles as industrialist were to be as volatile and as irresistible as his process. Almost all British pig iron has phosphorus in it, he discovered, and phosphoric pig iron in a Bessemer converter makes bad steel. When this came out, Bessemer said that by chance he had used a rare phosphorus-free pig iron from Monmouthshire in his experiments. Next he found a man named Mushet, who could remove phosphorus from a batch of molten pig iron by adding a manganese alloy called spiegeleisen. Bessemer began adding the stuff to his mixture in 1857, and rolled over Mushet when two of his trustees neglected to keep up the patent by paying the stamp taxes. Then Bessemer was able to continue using spiegeleisen without acknowledging—or paying—Mushet.

Between Bessemer and history stood only one more hitch, in the person of a Kentucky Irishman named William Kelly, who had endured at least one psychological examination because he insisted on blowing air through molten pig iron— and who claimed to have done so eight years before Bessemer tried it. Kelly further claimed that he had hired an English workman in those early days, a man whom he later recognized in photographs of the celebrated Sir Henry Bessemer. At first Kelly's suit was successful. The American commissioner of patents awarded him the patent on the process, upholding his well-documented case in 1857. In the meantime, though, Bessemer was furiously marketing his steel, and identifying the Bessemer name with the process itself. Kelly's claim was finally extinguished a few years later when he attempted to renew his patent. The American steel and railroad interests stood with Bessemer, to whom they would have to pay no U.S. royalties. By 1900, the Bessemer Process was the historical term, and Bessemer himself a multimillionaire. That year his company poured a quarter-million miles of railroad track out of its converters.

But in one respect, 1880 marked the end of the first era of the Iron Age. In that year, after more than three thousand years of smelting and forging iron, human beings at last produced out of Bessemer Converters a pure nickel-alloyed steel comparable in quality to the celestial iron, forged in the stars, which had fallen from the skies forever into the dissolving bath of the earth's atmosphere.

CHAPTER 10

The Catfish

ook at that," Harry said in satisfaction, as we stood around the configuration of cut springs, I-beams, and rock on the floor of the welding shop. I could tell he was beginning to see the completed catapult, but he wasn't directing our attention to any particular aspect of the springs and stone. As we were all already looking, his remark wasn't even imperative. Harry had a habit of saying "look at that" whenever he saw something he liked, as a kind of pure remark, a re-marking, an illumination of the moment. I know that Harry says "look at that" even when he is alone. And it's a habit he has conveyed to his wife and child—in fact, the phrase became his daughter's first words. For a while it was the only thing Julia said when she wasn't screaming. I'd carry her around and watch as she would direct her new dominion with the phrase, delightedly pointing and saying, "Look at that," as if she were creating the world, instead of merely finding it.

After he'd taken in the catapult by saying "look at that" a couple of times, Harry began to troubleshoot the device by imagining its future. We had to imagine a worst-case scenerio, he said, before we could engineer the catapult properly in the present. So Harry was imagining the springs—or the wings, as we began to call them—failing in the worst possible way, and maiming or killing us in the process. We would be putting the springs under thousands of pounds of pressure, he figured, more stress than a truck would ever put on them. We'd be bending them nearly into semicircles—no truck did that. And all that power would be aching for a way to release itself, to find some tiny flaw or brittleness in the steel, and shatter itself there. Harry said spring steel would explode like glass. He picked up the spring assembly and rubbed it with his palm. "Needles and shrapnel," he said.

Surely, I argued, the cable we'd be using as a bowstring would break first, and relieve the strain. Maybe, said Harry. But in that case, I didn't even want to know, he said, about broken cable. Broken cable could cut your head off, he said. It was true—I did not want to consider the possibility of such violence, but I had no choice in the matter, as David seized upon the moment to tell a story.

"That's what I've been trying to tell you," David said. "Broken cable is sharper than any razor." It seemed that David had been sailing one time, out on the bay, when he had witnessed the violence of broken cable. It was a rough day on the water, the wind beating the surface flat, the boat heeled way over, the water up over the gunwales. And as the crew pulled her through the wind to come about, a gust caught the momentarily slack sail, and broke a guy wire off at the deck. Lashed to the loose sail, the cable whipped and snapped over their heads, and the crew yelled warnings to the frightened passengers. But one Oriental woman didn't hear. She stood out on the bow in the roar of the wind, and when she turned around to attend to the commotion on the boat, the cable whipped across her body and slashed her diagonally from shoulder to waist. David made the gesture of the slash across his own body. He paused to let the impression of his story take, and then said, "Let's go upstairs."

I wanted to know more about David's story, which seemed too aimed to be true, for one thing. But I didn't ask. It didn't seem polite somehow to wring further details from him, so I never found out what happened to this dreamy Asian woman, up there on the bow above the wind-whipped water with her terrible cut. David evidently felt that the story was over, having made his point about the danger of broken cable. He'd created his effect, and he was sort of the boss that evening. If David wanted to tell a wild story and then go upstairs and take a break, that's what we'd do. It was his shop, after all.

Beyond the gold cars at the back of the shop was a steel staircase, narrow and without a rail. Beneath it was the table where the welders ate, and on the table amid the litter of old lunches was a pinup magazine of the old-fashioned sort— women in bikinis. We stopped to look. The bikinis were just straps, really, around their big breasts and round buttocks, and on their faces they wore expressions of festivity or mock surprise. David said that a girlfriend of one of the welders had visited the shop. She'd looked around and said, "What kind of a shop is this? There's no pinups!" So she had gone out and bought the magazine. Actually I believed this story, because the women weren't nude, first of all, and secondly because the magazine still looked crisp and new. It looked like nobody had even bothered to open it before Harry and I pawed through it. We didn't care.

We went up the steel staircase, and found Harry's sister and David's wife, Sandra, in the center of this relatively uncluttered upper shop, working on some kind of steel construction at a large workbench. She and her husband made one-of-a-kind things—decorative steel tables and headboards, techno-funk stereo speakers and the like, which were then purchased by clients. Sometimes Harry painted these things for them. Sandra was an ace at business and once had given me good advice about bidding on a project. She also had a good hand at making things. Susan said that Sandra made her own wedding dress out of white leather, cutting it up into a pattern like lace.

Sandra always looked striking, and she did even at the workbench, with her black pants and blouse covered by overalls and her hair pulled back to stay out of her eyes as she worked. She was using a small hand grinder to burnish the surface of a pair of semicircular steel rockers welded to this object, which I learned was a table. As long as I looked at this thing, I couldn't figure out which would be the upper surface of the table, or whether, as I had first assumed, the table was made to sit on the rockers for legs. Sandra scowled as she pushed the grinder into the surface, working shiny scallops into the steel. She kept burnishing it for a while when we came in, then put down her work and began to talk to her brother and her husband.

She spoke to Harry mostly, in the direct way that a lifelong relationship affords. He had been to see their father, and Sandra wanted to get the details. How was the old man doing? What was he driving these days? Did he seem to have a girl-friend? Did he ask about her? Harry answered these questions in a how-should-I-know? manner, noncommittally and without elaboration, and Sandra seemed somewhat annoyed, excluded from the two-man society of her father and her brother.

David chimed in once in a while; I didn't say anything. Harry's father was rather legendary to me—in my years of being Harry's friend, I'd met his dad only once, though he did business in my neighborhood. I knew many stories about him. I knew he drove a Porsche and an El Camino, both black. Sometimes I would look for his car on the streets. Later, when I did meet him, he was wearing one of those British driving caps, and was a sport, it turned out. He surprised us by show-ing up at the house one day with a gift for Julia. As soon as he decided he liked me, he began—for no reason—to call me Jake. He said that he was in the pants business and that he pulled down a couple of thousand a week.

David broke into the talk to tell me another story which seemed not to have anything to do with the discussion of Harry's father, but rather to emerge from some previous chan-nel of thought. When he lived in New York, David said, he had a neighbor who was an artist, but who wasn't doing very well in his art career. The guy never got any attention. So he decided to get a grant by proposing that he was an Artist in Seclusion. This impressed some big foundation guys, even though the guy was living in the heart of New York City, just two doors down from David. The guy actually got the grant, said David, to be an Artist in Seclusion. "I understand it, I understand it," he said. "Give me a fucking break, but I understand it." I nodded like I wanted a break, too.

Harry and Sandra had continued to talk about their father, and I felt called upon to respond to David's story with one of my own, to pull a related story from my story collection, the way one does in new situations. But I couldn't. I was sapped from the tension of working with the torch, and felt challenged somehow by David's story in a way that seemed too heavy to answer at the moment. Normally if one doesn't have another story about feeding apples to horses, for instance, one can tell one about throwing apples at trains, but somehow my brain wouldn't perform this dreamy leap, and as David looked at me, I just sat in silence, feeling more and more alien. Finally I suggested that I go and get Chinese food for everybody.

None of them had eaten dinner, so they liked the idea. David said that there was a place just up the street. So I left them there with their familiar conversation, went downstairs, and made my way out of the shop, past the dismembered ghost of a catapult and out the orange door to the street.

Outside it was cold. The wind was blowing hard, coming in off the ocean with that sea smell in it, and driving low clouds against the upper windows of the big Holiday Inn across Van Ness. The glimmering light under the western clouds was doing its last fade, and the low clouds, already reflecting street light, were descending and becoming fog. I zipped up my jacket and walked up Pine Street into the wind.

I was feeling a little down, the way I get sometimes when I

work with guys who work with their hands. I never took shop in school; I never worked construction. Two years before I had been one of a group of artists who had installed a new gallery space in an old maritime-parts factory. We had to clear the space of machinery and junk, put up skeletal walls of two-by-fours, wire the place, hang Sheetrock, and paint, and knowing nothing about carpentry, I did what I was told and volunteered to be the gofer there too, making beer runs as the other guys stood around with their arms crossed, gazing up at the unfinished walls and saying arcane carpenter things. Working on the project, I had tried my best not to injure myself, without much success. I am usually the guy who hits his thumb with the hammer, or carefully and with great effort puts something permanently in the wrong place, or lights his hair on fire.

So I went for Chinese food. I found the place on the corner that David had mentioned, and went in, out of the cold wind and into the smell of ginger and hot oil. The place had just a few tables in the back, behind the takeout counter. I decided not to look at the menu, but just to order a kind of lowest-common-denominator Chinese meal—sweet and sour pork, fried rice, one Coke, three beers. I had to choose between appearing unimaginative or ordering interesting food and taking the chance that everyone back at the shop would hate my selections from the menu, and it was no contest—I would be dull and safe. But as I wasn't ordering from a menu, and the woman behind the counter didn't understand my English, I had some trouble making myself clear. As soon as I said, "Sweet and Sour Pork," she said "Number?" I looked through the menu until I found the number. It was number 14. Fried rice was number 47. After I ordered, I was given a slip with my number on it, and I sat down in one of the three white welded-wire chairs in the waiting area, next to a woman who was waiting for her order. She had on a suit, a purple silk blouse, and running shoes.

"Have you tried the catfish here?" she asked me. I said no.

She gestured to the foaming tank next to her, that made the fourth side of the little square in which we sat. "They're alive," she said.

They were, sort of. In the big Plexiglas tank, two or three of the fish swam around, looming in and out of the cloudy water, their long whiskers wavering. The rest of the other dark creatures just hovered disconsolately at the bottom of the murky pen, opening and closing their mouths, as if attempting to speak, then giving up.

"I just got sweet and sour pork," I said to the woman. "The old standby." That seemed to kill our conversation for the time being, anyway. And after a while the woman had her number called, and she picked up her food—catfish I assumed—and left me there, watching the fish in the tank.

I sat there waiting for my food, feeling worse and worse for these fish. Where had they come from? The Yangtze? The Sacramento Delta? Had they ever even had a wild home, some silty place beneath the eddies where they could skim the bottom? How had they arrived in this predicament, trapped and condemned in a Plexiglas pen in a restaurant in the middle of a city? Had they been captured? Had they been netted or hooked? Or had they come from some sort of wet farm, a crop, raised in something not unlike this pen? Not knowing the difference might be preferable, I thought. Or maybe fish had no memories in any case.

As I watched, one of the swimming fish seemed to become desperate. He was a long black creature, big as a baguette, and his drooping whiskers made him look a little like a gambler or an outlaw. All at once he leapt for the top of the tank. He broke the surface of the water, hit the tank's clear plastic roof and fell back, thrashing. He leapt again, and a third time, each time thumping hard against the roof and falling with a splash among his disturbed brethren. Then he was still. He lay gasping with the others on the bottom. He had muddied the water with his thrashing and I could hardly see the fish anymore— just the dark voids of their mouths opening and closing in the gloom. Then my number was called and I got my order and paid and took the food back to the shop.

The three of them were still there at the workbench on the upper floor, talking. We passed around the paper cartons. In one was a surprise. The sweet and sour pork that I had ordered by number turned out to be prawns with baby corn in oyster sauce. Maybe I got number 40 instead of number 14. But everyone was hungry, and Sandra turned out to be fond of baby corn. I didn't confess; I just took the credit and ate.

After dinner we sat around the workbench, everybody but Harry smoking. I bummed a cigarette from Sandra, and remembered a time before she was married, when we had spent an evening together at Harry's. That evening we had been relatively formal, as we hadn't known each other well. And afterward we left the building at the same time and drove in separate cars back to San Francisco. I followed her up the freeway ramp and onto 580. In a mile or so I pulled up alongside her as we came to the tollbooth lanes for the bridge and lined up to pay. We waved to each other as we entered our gates, each of us in the relative stillness of our separate cars. That moment had remained our friendliest, I thought, made possible somehow because of the distance between the cars, the layers of glass, and the space of pavement between us— all that inviolate insularity of the freeway.

Sandra asked Harry if we could give David a ride home, then took David's car and left. We men went back down into the welding shop to finish up. We trimmed the I-beams with the suicide saw, and used a big drill press to put holes through them, so we could bolt them to the springs. David said he didn't have anything that we could drill the springs with—that was specialized work, he said. So by the end of the night, we'd finished what we could, and we loaded our booty into the back of Harry's truck. David shut off the lights inside the shop, emerged from the utter blackness with his key, and locked up.

Then he squeezed into the cab with Harry and me. We drove off happily, having done some work. David gave directions, recommending the timed lights on Pine Street over any in the city, and we dropped him off at his house in Pacific Heights.

"Well," said Harry, when we were finally alone in the truck, "was that awesome or what?"

"Definitely awesome, Harry," I said. "Fun even."

It was late by then. The streets were vacant and fog billowed under the lamps. Eager to get home, Harry drove fast, cornering the truck hard. Behind us our catapult stone rolled around in the dark bed, hopping over the shifting steel and banging against the sides, as if it wanted out.

CHAPTER 11

The Warwolf

Dolores Park in San Francisco's Mission District is a steep rectangle, a grass plot falling diagonally across the side of a ridge beneath Twin Peaks. A sidewalk encloses the park, not one foot of it level, and walking up it one day, I saw two men fight on the hillside. They were young men, like rams on the slope, a woman looking on from above. The larger man had longer hair and a pale blue Oxford shirt. He lost the fight.

At first, they were taunting each other in Spanish and one of the insults proved too much to bear. The smaller man struck the other with a cuff, not a fist, and the other seemed amazed and infuriated. He hit the small man hard a couple of times with his fist, but this nasty, impetuous shrimp bore up under it, then kicked him back hard in the hip.

The smaller guy did a couple of things right, besides kicking. He used gravity and his opponent's own weight against him— staying uphill and concentrating on putting the other man on the ground somehow. Also he humiliated his opponent by ripping his pale blue Oxford shirt, not once but several times, finally leaving just pieces of it, tatters, on the guy.

At the turning point in the fight, the larger guy was getting up off the ground, pacified, gesturing in astonishment to the remains of his shirt, which was just the cuffs at that point, those and a scrap hanging on the waistband of his jeans, and the smaller guy took the moment of this gesture—both hands open to the body to display it—to kick the big guy in the naked ribs that he was displaying. That sent him down again, and when he got up, he just walked off, left the park. His opponent followed him, kicked him in the butt once for good measure, then went back up the hill to the woman. Shocked and disconsolate, the big guy tried to act normal when he got to the sidewalk and walked past me; he with no shirt left except the baby blue cuffs around his wrists. I wondered if he had a gun at home.

Approaching Scotland from the east or south, one is naturally funneled into the Forth River valley, verdant lowlands where the river meanders, gathering the streams called waters in Scotland that run out of the Highlands. Just where the Forth becomes too shallow to navigate, a natural tower of black stone rises from the meadows of its southern bank. The core of an equatorial volcano, the rock was formed three hundred million years ago, a period when even the dinosaurs lay inconceivably far in the future. The volcano cooled and hardened, and for ensuing eons it drifted northward with the continent, until it grew heather and scotch broom and was slathered with ice. When the earth warmed and the glaciers retreated, the melting ice tore away the earth on the volcano, leaving its black stone core standing like the prow of a ship in the silty lowland, a sloping tail like the ship's wake drifting down from the top in the direction of the glacier's retreat.

This black rock pinnacle at Stirling was the northern reach for tribe after tribe of invaders. Beyond it, the Highlands offered an advantage to the individual warrior, to the guerrilla. So successive generations of invaders occupied the lowlands in the east and south, while in the mountains wild marauders passed on their heritage of revenge. Rome found one limit of its empire here. With their roads and catapults, the Romans came into the Forth Valley, looked up at the Highlands, and left them unconquered. Gibbon writes that the legions turned with contempt from the gloomy hills and cold heaths where naked barbarians chased the deer. A few miles to the south, the shivering legions erected a turf rampart on foundations of stone to keep out those wild Caledonians. Ultimately Rome withdrew from the island altogether, leaving a people so technologically benighted that for centuries the Roman feats of engineering would be supposed to be the work of giants.

In the Middle Ages, the Scottish terrain posed its old difficulties to a new set of invaders, and the English could only take what the Romans had taken, conquering the plain and seizing the rock at Stirling, which they fortified as a command post against the local resistance—fierce men in plaid who had again taken refuge in the mountains across the river. But a fortress is only as good as the power beyond its walls. Without such control over the surrounding territory, a high, fortified place, exposed as it is imposing, becomes a self-made prison. So it happened in 1297, when the Scottish rebels under William Wallace mounted a counteroffensive against the English invaders, and the English lost their castle not at its walls, but a mile away on the valley floor near the river at Cambuskenneth.

Wallace, a commoner, had taken advantage of infighting among the Scottish nobility to gather his own revenging army, rubes with homemade swords whom the English landlords did not at first take seriously. Scottish lore holds that the English exchequer ordered the battle casually, in order to curtail the ongoing expense of maintaining the troops for some eventual defense. In any case, these medieval English troops came down off their rock, rode over the narrow bridge at the River Forth and into enemy territory. There the enemy was ready for them, and the English were promptly cut off by the Highlanders, who raged out of their wooded hideout on the Abbey Craig, secured the bridge, and slaughtered the hundreds of luckless Englishmen who had ventured across. Routed, the remains of the

English force fled south, abandoning their castle on the basalt promontory, which the Scots took over, dancing their jigs on the battlements.

The English king Edward I was an old man who had learned at age twenty-four the virtue of nailing things down, of leaving nothing to nature or chance. Then an impetuous prince, he had taken a command beneath his father's banner to put down an uprising of the barons and the Londoners. The barons in England had risen against the crown, urged to rebellion by the weak rule and high taxes of Henry III, Edward's father. A petty spendthrift, they called him. The Londoners joined them when the king arbitrarily removed their elected mayor and sheriffs, replacing them with his puppets. These Londoners had helped precipitate the rebellion. They committed an outrage against the queen. When Edward's mother, Eleanor, was passing in her royal barge beneath London Bridge, a mob threw stones and garbage upon Her Highness. It was an insult her son would not forget.

A year later the revolt was in full swing and the forces of the king found themselves arrayed against the rebels at Lewes Castle in Sussex. The prince commanded troops beneath the castle walls on his father's right, and when the enemy came into view, Edward recognized them at once as Londoners. Swinging his sword in a sentimental fury, he charged the ranks of these louts, mother-haters, stone-throwers, and when he looked up from the carnage, he found himself and his troops cut off from the main force of his father's army. Thus depleted, the king's troops were defeated by the rebels, and the king himself captured. Edward had to be given up as a hostage to secure his father's freedom.

The next year Edward escaped from prison, a changed man. He regrouped his father's forces, and smashed the rebel coup with a strategic, determined, and multipronged campaign. Edward returned his bitter and foolish father to the throne, and departed for the Crusades, where he would study the science of war. He routed the infidels from Nazareth and was terribly wounded afterward, attacked in his chambers by an emissary with a poisoned dagger. As he recovered, his father died, and he returned home to be crowned.

As king, he would not fall prey to flamboyance. His one instant of youthful passion and failure, his mistaken assumption that one spirited dash might rout the forces of chaos, he put behind him. In the course of his regency, Edward subdued the flamboyant, disorganized Welsh with the slow machinery of the siege; expelled the Jews from England and seized their property; and unified the nation with long-term legal maneuvering, convening the prototypical Parliament not so much to give the people a voice, but to place more effectively the yoke of government upon them. But his final and most grinding effort—a failure—would be his campaign to conquer Scotland.

As Edward was modernizing the census for tax collection in the south, Scotland was burning with flamboyance. On a dark night in 1286, after much food and drink in his castle at Edinburgh, Scotland's king (and Edward's cousin) Alexander III conceived an urge to see his wife, who was residing across the Firth at Kinghorn. It was a bad night to travel, and the king was warned, as in a fairy tale, three times—by his courtier, by the gatekeeper, by the boatman—but to no avail. His Highness was urgent. On the far shore he took to horse and vanished into the fog. At dawn, both he and the horse were found with their necks broken, dead on the beach below the sea cliff, having apparently ridden off the precipice in the dark. The ensuing struggle for Scotland's empty throne would challenge England's rule and open decades of vast, seasonal slaughter.

And so as King Edward reached retirement age, along came these chaotic Scots with their interminable uprisings. The king had to ride into Scotland at the head of his army several times, seeming each time to defeat the rebels, subduing the Scottish nobility most decisively at Falkirk in 1298. There, the Scottish nobles had set out decorous squads of pikemen called schiltrons, an excellent defense against knights on horseback, but most idiotically vulnerable to Edward's squads of English longbowmen, who cut them down from long distance. The schiltrons held their ranks and fell as one, an unsporting death and a complete victory. After Falkirk, Scotland was subdued, it seemed. Edward had a brief respite of bureaucracy in London. But again the rebel submission was momentary, and within two years Stirling Castle was back in Scottish hands.

Ponderously and with heavy armament, Edward bore down on Scotland once more. On Midsummer's Day in 1304, he convened his government and assembled his army—a mobile city—at York. In his train were as many as three thousand knights with their grooms, squires, attendants, and household regalia, and behind them perhaps a hundred thousand longbowmen and infantry, or at least so say the chronicles. Going north to war as well was most of the king's administration, the mechanism of the government—nobility of the household, clerks, and officers, chief among them the accountants of the Exchequer. And following were a Chaucerian aggregation of representatives of the rest of medieval society, parasitically attached to this military host—priests and cooks, laundresses and chroniclers and whores. Edward commissioned a corps of gravediggers, paid, as in the parable of the Vineyard, at a penny a day. And among this rout were the engineers, joiners and carpenters and foremen, overseeing transport of the parts of the king's large engines of war, to be assembled on the site of battle. Ships towed portable bridges offshore opposite the procession, bridges that would span rivers in the north, especially the Forth, where the rebels at Stirling Castle commanded the prime crossing.

Beneath hundreds of fluttering forked pennants, Edward's vast host marched north. At its head, in his helmet and spurs, the king carried himself well for a man of sixty-five, ancient by medieval standards. On his doublet and shield, and on the armored cloak of his small horse, he bore the triple lions of his forebear Richard the Lion-Hearted. The helmet looked like a welder's mask with a crown attached. On its hinged faceplate was a large Gothic cross, slotted on its horizontals for his eyes. His chain-mail suit—woven rings of precious steel—extended from his wrists to his ankles, and over it he wore a flowing tunic, covered with brocade, which flapped at his thighs as he rode. His sword, which had slain Saracens at the Dead Sea, hung from a belt at his waist, a crucifix carved into its hilt.

His army's pilgrimage into Scotland was slow and devouring. He met no organized resistance. The towns en route were smashed and burned, the people killed, enslaved, or exiled, their possessions seized. The host swept northward like an enormous antibody, a miles-wide macrophage, grinding up the narrow muddy lanes through the heath, across the remains of Hadrian's Wall, built by the Romans more than a thousand years before. By the end of the summer, Edward had crossed the Firth of Forth and set up headquarters at Dumferlaine, the place where, nearly twenty years earlier, his cousin's broken body had been carried after he had ridden off the sea cliff. Edward passed the Scottish winter in the abbey there, and when spring arrived, the Scottish nobility sued for peace.

This time both Edward and the Scots knew that official submission would not suffice. Had the chivalric code of the feudal order still held here, Edward would have secured the country two years earlier. But Edward's ravages against the population, for one thing, had kindled a new kind of uprising, a nationalistic and democratic rebellion that answered to no earl or baron. So despite the surrender, Edward took care to mop up the resistance. A hundred or so ordinary Scots led by a plebeian commander, William Olifard, continued to hold Stirling Castle, the country's essential chokepoint. At Stirling, Edward would bear down.

So the English host turned west and marched up the Forth River valley. The defenders of the castle must have seen the slow horde coming for days before it arrived, and have recognized instantly their hopeless situation. Implacably this tide of human beings swept in around them, around the tower of rock beneath their walls. At midsummer, a year after he had begun his campaign, Edward's force had ringed the citadel, and the siege had begun.

The king could afford to be patient, and there was in this siege, as in other hopeless cases such as the Roman siege at Masada, an atmosphere of performance, of cool theoretical demonstration, rather than the heated air of battle. The purpose of the siege, this late in a foregone conflict, was not simply to capture the castle. The besiegers would demonstrate the fate of rebels against the crown by exhibiting the might and technological accomplishment of England and her king. Edward seems to have savored the siege at Stirling, performing it as a succession of flourishes, saving the grand gesture for the last, as in a bullfight. For that last flourish, Edward would deploy a huge catapult, which the king called Ludgar, his Warwolf.

For several weeks, there was an ongoing, low-intensity battle as the English dug in and began construction. The king's bowmen skirmished with the defenders. Edward then set up his own small catapults, thirteen of them, and began a regular bombardment of the walls. He ordered materials for Greek fire—a fearful liquid packed into hollow clay projectiles, its blaze unquenchable by water. Then Edward also commanded his crew of carpenters and engineers—perhaps fifty men with five foremen—to begin assembling the Warwolf.

Edward showed himself for the benefit of his troops, riding around the battlements within range of arrows from the walls, exposing himself to fire. The official chronicle says that a bolt fired from a springald, a sort of large crossbow, narrowly missed him, plunging into the leather of his saddle without harming him or his horse. A huge stone, flung from a catapult on the walls, bounced between his horse's legs, or so says the chronicle.

The battle was intimate and quiet by modern standards, a

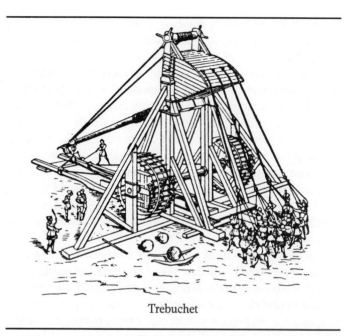

Trebuchet

Illustration from *The Crossbow*, by Sir Ralph Payne-Galway,
The Holland Press, Ltd., London, 1903

silence in which one could hear incoming arrows cutting the crisp air. The combatants could watch each other easily, and individual opponents came to know each other. The king's forces taunted their captives by announcing their preparations as they made them. The defenders shouted insults from the walls. They called Edward "Longshanks," says the chronicle, because of his long legs. One supposes they called him worse—they had a long time to consider insults. The Warwolf's parts had to be transported overland, then hauled up the steep slope by crews with ropes, rollers, and pulleys, and erected below the walls of the castle. From the walls, the Scots could comment on their enemy's slow progress on the machine.

* * *

The Warwolf was probably a gigantic trebuchet, or sling engine, working by a counterweight, like a grandfather clock. Later illustrations of such weapons show a skeletal tower, fronted with eaves to protect the workers from showers of arrows and stones, and braced diagonally, so that the machine would not fly apart upon firing. The tower bore as its uppermost rafter a thick round beam upon which hinged the entire trunk of a huge pine, which extended diagonally behind the tower to the ground. Above, this tree trunk was bolted and bound to a room-sized wooden box, a counterweight filled with several tons of earth and stones. Below, workers lashed the sling itself to the trunk's tip. The sling's pouch held stones of two hundredweight (two hundred pounds), balls of rock sixteen inches in diameter.

Gangs of men worked huge hamster wheels on either side of the tower, treadmills that raised the counterweight. When it was ready—the heavy box raised and pinned, the crew removed to one side—the commander would drop a raised baton, and the triggerman would pull a rope, freeing the weight. Then the trunk would lash its whip, and the sling would release its stone. After the shot, the whole cumbersome enormous mechanism would heave back and forth until its antlike crew could steady it with ropes, climb back into their treadmill wheels, roll another heavy stone into the sling's pouch, and begin the laborious business of cocking the catapult again.

But the defenders of Stirling Castle didn't wait for the actual firing of the Warwolf. Just witnessing its construction, its enormous beams, the hammerblows, the winching of its parts into place—to say nothing of the sheer tenacity of the besiegers—they were convinced of their doom. On the day the Warwolf was completed, they capitulated. They ran up the white flag and sent an embassy through the gates. They surrendered absolutely, much to Edward's disappointment.

For the king had fallen under the spell of the great catapult, too, by that point, succumbing as well to the slow suspense of its construction. He had become enamored of the technology, and beside this grand demonstration of his power, the capture of the castle had become a technicality. This weapon had taken months to ship, weeks of effort to build. It had been expensive. The best technical minds in the kingdom had labored on its plans, the king himself assisting in the design. After all this, Edward was not about to allow a few surrendering Scots to interfere with its performance. The Warwolf, its fierce persona enhanced by the slow crescendo of its building, would be fired, irregardless. So the king ordered the surrendering Scots back into the castle, to defend themselves against his might and majesty as best they could. Nonplussed, the rebels went back inside and bolted the gate. Then the great catapult was loaded and fired.

The chronicle of the siege records that the Warwolf's first stroke broke down the castle's entire wall. Then the heavy stones must have fallen into the masonry walls of the buildings within the perimeter, collapsing them. Then Greek fire was shot into the wreckage, setting it aflame. Then Edward strode through the broken gate, triumphant amid the rubble, ordering the surviving defenders executed or led away in chains. Stirling Castle was again an English citadel.

But neither this possession of the castle, nor its possessor, were fated to last. Within two years, the new Scottish champion, Robert the Bruce, had taken the reins of the ongoing Scottish rebellion. Once Edward's ally—it was he who supervised the transportation of the Warwolf's parts into the north—Robert assassinated his chief rival among the Scots, stabbing him to death in a chapel in Dumfries. He then began his own war against the English.

So at sixty-eight, Edward again hauled himself into the saddle, returning to Scotland to do the whole bloody job over. On his way north, he fell ill with dysentery, and, as his attendants lifted him to his place at the table, died. Dogged even in death, Edward had directed that his remains accompany the English forces, now under the command of his son, Edward II, until Scotland was conquered. Had his wish been honored, his bones would have remained in transit for two and a half centuries. As it was, his body was returned to Westminster Abbey for burial beneath a stone inscribed in Latin with the words, "Here lies Edward the First, The Hammer of the Scots, 1307. Keep Troth."

Atop its black pinnacle today, Stirling Castle still commands the countryside. The visitors clutch their coats and take in the view from the windy battlements: the valley floor, hundreds of feet below, the wild hills rising in the west. In October, the sky is often lidded, pierced here and there by bolts of white light. Flocks of crows roost in the black cliffs beneath the towers. Late in the day, as if at a signal, they depart, spiraling over the rock in a ragged mass, raucously crying.

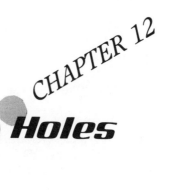

CHAPTER 12

Holes

Then it was October—our deadline in early December beginning to loom before us—and all we had done were those springs. Worse, our bluff was getting called. I went to a group opening of enormous paintings, and was startled when someone asked me about the catapult. It was going just fine, I said, heading off as if to mingle. How did she know about it? I thought. Word traveled so fast in this pioneer port.

Worse yet, an environmental sculptor cornered me when I told him about the project. "We're doing the welding now," I told him.

"Oh really," he said. "What are you welding?"

"Actually we're just cutting out the pieces," I said.

"I built a catapult once," he said, "a kind of big crossbow."

"Fabulous," I said, wondering with some desperation how I might get out of hearing about it. He'd built his catapult to shoot lengths of conduit fitted with flares, he said, and he had invited his friends to watch him shoot it. They'd taken it up to the top of some incredible quarry.

"How did it work?" I said.

"Well, actually," said the sculptor, "it didn't." The springy conduit absorbed too much of the stress of the shot, he said. It sounded to me like the conduit simply fell quivering into the quarry. This guy wasn't so bad, though, I thought. At least he had failed.

"Gee, too bad," I said. "We won't use any conduit, in that case."

Then the next day the press officer from the art center called me up. She was preparing a press release, she said. It's barely October, I said. She told me that the magazines needed to know two months in advance. What did our catapult look like? she asked. At that point the catapult looked like zilch, like a few chunks of heavy raw steel, just lying there on Harry's porch, the raw spots going red with rust. I told her I'd get back to her. I thought about just giving her Harry's phone number, but decided not to. Harry would kill me.

But that evening I went over to Harry's to prompt him into taking a further step on the catapult. At that point we were simply looking at the stuff. Harry's method of engineering required long periods of silent staring. We had to tease the progress out of the tangle of ramifications, as if at chess, but it felt more like we were attempting, by silent prayer, to will the thing to life. We'd gotten out two folding chairs, and had found some C-clamps and clamped the springs to their mounts, setting the loose parts in formation, but the steel hung disjointedly together, lifeless. And Harry just sat there, staring abstractedly at the stuff.

"What's conduit?" I asked him, after we'd stared for a while.

"There's all kinds of conduits," said Harry. "A sewer pipe is a conduit."

Somebody had shot conduit out of a catapult, I told him. "Probably some kind of electrical conduit," said Harry. It didn't work too well, I told him.

"No wonder," said Harry. "Even when you throw it, it wiggles in the air." This said, he went back to looking at the springs. But he looked a little more interested, I thought, less like he was just doing it for me, at any rate.

So I bore down on him further. The art center's doing a press release, I said. I reminded him that once we accepted the money from them, we also accepted a schedule for completing the project. We had to have a successful shoot in early December, and just after that, a public presentation. Harry said I didn't need to tell him any of that. The public presentation was my business.

"What are you going to say about this, anyway?" he added. "Have you made up any findings yet?"

"Harry," I said, "say something, do something. I'll have findings."

"What am I?" he said. "Some kind of character or something?"

I tried to reassure him and get him off the subject. "Don't worry, Harry," I said, "I'll be perfectly objective."

"I don't like the sound of that," said Harry. But after staring a while longer he added, "Eccentric pulleys. I think we might need eccentric pulleys." I had no idea what he was talking about, but I said nothing, hoping this clue would lead us somewhere, out of the thicket. I knew he was worrying about the project, and I was glad. He was beginning to share my sense that some exposure loomed up in our near future.

I was secretly afraid that we had jinxed things by going public. As a lark, the project stood a chance. But by accepting the money and making its completion contractual, we had defied the boy gods, somehow. We would slip off schedule, do something reckless, mess up, get angry at each other, and be battered by the mounting pressure, until one of us quit the project. And the thing would not be built by one of us alone. For my own obscure reasons, I was driving the project. But I was no engineer. I couldn't make it work by myself. Harry had the hands for things.

"Did you say something about pulleys?" I said.

Harry said, "I was thinking of a way to boost the power of this bow. All the new bows have pulleys on their tips."

Gingerly, I kept him talking. The bowstring passed through

these pulleys, as I understood it, so that a much stronger bow could still be easily cocked. Further, the holes in these pulleys were off center, eccentric, so that the bowstring would accelerate when released, as the pulley rolled off its high side.

It made my head hurt to think about it. All I wanted to do was throw some rocks. This pulley business was like quantum physics to me. "Harry," I said finally, "do we really need these things?"

"I want this thing to work," he said. "The springs may not be enough." He was afraid that we'd get up there in the Headlands with something like a catapult, and the whole project would unroll in front of an audience like a limp hose—a phallic catastrophe, merely plunking out its stones in limp arcs, a complete flop. Harry just wanted the thing to work, that was it.

I looked down at the springs. They looked like some kind of huge bear trap on the deck, and seemed enormously strong, strong enough to launch a truck, two trucks. I didn't say anything, though. Harry was rolling, and I just wanted to keep him in motion. His direction seemed a minor matter compared to that.

"OK, Harry, you're the engineer," I said. "If we have to get these pulleys, then what's the first step we have to take—the first little step."

"Have to drill some holes," Harry said. I wanted to hug him. Holes I could manage. "Where?" I asked.

For the first time in what seemed like weeks, we got out of the chairs and touched the steel. "Here, here, and here," said Harry, pointing to the ends and the middle of one of the bows. I wanted to get out a drill then and there and begin the work, but Harry reminded me that we couldn't just drill the springs ourselves—we'd have to have special bits and a really slow drill press.

"This stuff is going to have to be professionally machined," he said. "We're going to have to find a machine shop to drill them. And we should get that inner leaf tapered, too."

* * *

The next morning right at eight o'clock, I called Precision Machining in West Oakland. I'd found them in the Yellow Pages, and as the phone rang I had the feeling that I got sometimes on the road, when I memorize a question in another language—"How do I find the nearest subway station?" or something. I always hoped to get a one-word answer, or a set of significant gestures, anything except a whole torrent of unintelligible Italian, say. Likewise I knew a few key terms for the machinist, and I hoped not to be asked any detailed questions.

A woman answered the phone. I told her I needed some high-grade steel drilled and cut. It was a small project. She said I'd have to talk to the manager, whose name was Art, appropriately enough. She put me on hold for a while, where I listened to Frank Sinatra singing "Young at Heart." "Think of all you'll derive just by being alive," he sang.

Then Art came on. "What kind of steel is it?" he asked. High-grade steel, I said. "Truck leaf springs."

"What's the grade number?" said Art, nailing me with a single shot. I had to admit I didn't know, exactly. He said I would have to bring it down there and let him see it. Fine, I said, I'd be down there later that morning.

I hung up, called Harry, and made him come with me. I didn't want to get into another fiasco like the one I'd undergone with the Men in Brown. So I picked up Harry at his shop, and we loaded the steel—still assembled with its C-clamps—into the hatchback of my Honda, and drove over to Precision Machining. Harry noticed something odd about the shop's windows.

"Plex," he said. "This guy's got serious bulletproof windows."

In the outer office, there was a friendly woman at a desk, and a low barrier with a little gate in it. Harry and I walked in, both of us with a catapult wing slung over one shoulder, where it was heavy enough to hurt. The woman announced us by phone and after a moment told us to go through the little gate and the rear door.

I was a little surprised to find that we stood, not in a steel shop or a warehouse, but in a well-appointed inner office. Art sat behind his big desk. With our raw weaponry on our shoulders, we introduced ourselves. I could tell Harry was less than comfortable in the office. He looked like we'd been sent to the principal. Art was a thin, muscular man in his fifties, tan and mostly bald beneath his crew cut. He had strong-looking jaws, and an animated, strategic sort of face. I felt him size us up in an instant.

"Is that Plex in the windows?" Harry said, just making conversation.

"Yessir," said Art. "Kevlar. We used to have a lot of trouble with breakage. That stuff has more than paid for itself by now," he added. Art's jocular, nonchalant way of talking made me nervous.

We were quiet a long second, and then I said, "We're doing a little project."

Harry said, "It's a catapult. We're making it as an art project. We got a grant to do it."

"A grant?" said Art. I looked incredulously at Harry, who had hardly said a word about the project for weeks and was suddenly bragging to this character about our source of funds.

I tried to gloss over the part about the grant. "We need to get these springs machined," I said. We lay the steel down carefully on his desk and indicated where the holes had to be drilled. Harry had also drawn a tapering line on the smaller spring to indicate where he wanted it cut. The spring had to be tapered, Harry had explained to me, so that it would assume the force of the other spring more gradually. Art seemed to understand all this, but after we'd finished explaining, he turned his palms up and said, "I just don't know."

Spring steel was tough stuff, he said. He could probably do it, but he couldn't give us an estimate on the cost of the work.

These holes would have to be drilled slowly, he said, very slowly—if they could be drilled at all.

"How much do you charge per hour?" Harry said.

"Forty dollars is standard," said Art.

One of the guys in the shop, Art added, had been a soldier of fortune. He'd probably enjoy taking this on. Why didn't we just leave the springs there, and they'd take it from here.

We said fine, left the springs, and got out of there. "Why did you tell him we got a grant?" I said to Harry as soon as we hit the sidewalk.

"I wanted to let him know that somebody else was paying us to do this," said Harry.

"Fabulous," I said. "Now he thinks we're loaded."

"And that shop was kind of expensive," Harry said.

"Oh hell," I said, "Let's not worry about it. How much can a few holes cost, anyway?"

A lot, it turned out. The next day I called Art to find out how much it would be. "Your springs are ready," he said.

When I got over there, the job was waiting for me. The leaves had nice neat holes in them. The inner two had been tapered and beveled. That would be two hundred dollars, Art said. These had been, apparently, extremely slow holes.

The bill seemed outrageous, but I didn't say anything. Maybe I was supposed to suffer in order to get this project under way, I thought. Besides, we had bought holes, and there was no returning holes. I wrote Art a check amounting to more than half of our remaining budget for the entire project. And for holes, I thought, for having nothing where something was before.

I took the steel and our precious new holes over to Harry's. I hadn't realized that, in getting this tempered steel, we had committed as well to the extra costs of working with it. You paid dearly to have it, and then you paid just as dearly not to have it, I thought as I pulled up in front of Harry's, where he turned out to be standing at the window in the living room. As I got out of the car, he came down the front steps, wanting to see the work. I opened the hatchback and showed him our newly machined springs. Then I told him about the bill.

"Wow," he said. "I sure hope we need those holes."

CHAPTER 13

The Beam

Though they were nothing, in and of themselves, the holes were calls for further action. Holes want to be filled, and we did what we could to fill them immediately. We drove to a hardware store and bought bolts. Harry knew about hardware stores—he bought tools for his work—and this place was down under the freeway near the Oakland docks. The inventory had enormous range, I thought—bins for bolts, for instance, that ran from eighth-inch machine screws to steel clubs for God-knows-what. And about a third of the way from the small end of this bolt scale, we found some three-inchers, a little thicker than pencils, to put through the new holes in the springs and the old holes in the I-beam mounts. We also got washers and nuts, and hurried back to Harry's, where we replaced our shifty C-clamps with tight new hardware. For the first time the springs stood on their own.

In the meantime Harry was anticipating ramifications. Worse yet, he was doing it out loud, telling me what we couldn't do before we did whatever. Filling the other holes on the springs required the pulleys. And we couldn't make the pulleys until we could cock the bow. We needed to see how far the bowstring would have to be pulled back, Harry explained. That distance would dictate the circumference of the pulleys.

Installing the bolts had quickened the entire enterprise, I felt. Building and firing the catapult wasn't going to proceed in a series of equivalent steps, but, as in a chain reaction, with each step radiating further steps, the insistence proportionately stronger, as if the finished catapult—or, rather, the flung stone itself—already existed somehow in the future, I thought, and had some sort of psychic gravitation that increased as the project neared its embodiment. As if it wanted to be whole.

In any case, I just wanted to do it, whatever it was, and Harry's explanations were intolerable to me, though I had to admit that they were necessary, descending from the demands that our stone—so much weight, so much mass—placed upon a successful firing. I didn't have to like them, though. In his allegorical prophecies, William Blake described humanity's fall in such terms, I remembered. According to Blake, you listened to Sir Isaac Newton and you ended up subject to gravity, a black hole, a squeezed core.

So when Harry began to tell me why we couldn't install the mounts, I covered my ears. "Harry. Harry," I said. "Just tell me what we do next."

"We can't install the mounts until we get a beam," said Harry. "I guess we could get a beam."

At first we had considered using a steel I-beam for a stock. We had even hefted one at David's to see how portable it was. Not at all, I decided immediately. So we had begun thinking about a wooden beam, a timber or something, about ten feet long, we figured. My impulse was simply to go to a lumber yard and buy one, preferably oak or maple, something blondish anyway. Harry thought this a disgusting suggestion. "We're not making a kitchen appliance here," he said. We happened to be in the kitchen of his house when he said it, and Susan walked through.

"Besides," said Harry, "maple trees are too small to get beams from."

"And anyway," Susan said, "an oak beam would cost a fortune. You should find some demolition place." They'd have lots of old beams, and cheap ones, she said.

"A demolition place, something like that, that's what I'm trying to say," Harry said.

Susan got the paper and found an ad for a salvage yard. "Specialists in Wood Beams," read the copy.

Life intervened for a while, for almost a week. It kept doing that during this project. Except during the most intense weeks, we'd be interrupted by our jobs, or by family stuff. This time I think Sara's sister came to town, and we spent three days showing her San Francisco. So I didn't get back to Harry's until the following Saturday. When I got there, he led me back through the house to the porch, and gestured at the catapult wings. They were bigger. Harry had done some work on his own during the week. He had gone back to David's and welded plates of steel decking onto the I-beams. The plates were wing mounts, Harry said. They were diamond plate, having that cross-grained diagonal nubbing on their surface, as on the ramps of car ferries, and Harry had drilled holes through them, so that we could bolt them onto the beam when we got it.

I felt both relieved and annoyed. On the one hand, Harry had finally done something on the project without my having to push him. On the other, he had hogged this step in the process, and it made me a little jealous. I was going to have to watch him, I thought as we went out to find a beam, to make sure he didn't run off with our catapult.

The place Susan had found for us, Aldo Demolition, was deep in Oakland. Harry and I drove down 880 in the truck, into the vast sunny ghetto, down along the railroad, underneath the freeway and the elevated BART tracks. The salvage yard stood on a street of tiny and funky California bungalows, near a boarded-up supermarket. A high wooden fence enclosed the yard, which was just across the street from the Tabernacle

Missionary Baptist Church, with its orange stained-glass windows and sign reading JESUS SAVES. Inside the demolition yard lay acres of stuff laid out in stacks, dominated by towering piles of wooden beams. In the middle of the yard was a little green house, closely surrounded by another fence, this one chain-link with concertina wire wound lavishly over its top. Inside that fence, a little front yard held only a narrow sidewalk and a single tree, a spruce, its trunk stripped bare of branches to the roofline of the little house, but bushy and green, oddly healthy-looking, above.

In the office a young, disheveled woman sat at a desk, her feet in pointed boots upon it, watching a soap opera on a dim TV. "It's just me," she said, as we peered in. No one else was around. We could look around the yard if we wanted. Her boss should be back soon.

Harry and I backed out of the office and wandered into the wasteland. A dirt road ran back into the yard, snaking between the stacks of beams. Around the first bend, we were confronted by two filthy dogs. They were big, grimy mutts, as matted as sheep, one brownish and one blackish. They stood up when they saw us, but they didn't bark. A bad sign, I thought.

Harry said, "You think she would have sent us back here if these dogs were nasty?"

"No telling," I said. I began greeting the dogs in a tone that might let them assume that we thought they were cute. It worked. They came over and fawned around, the blackish one licking Harry's hand. "They're friendly," he said.

So we continued into the yard, moving between the mammoth stacks of wood, some of them looking precarious, and followed by the dogs, who, having greeted us, seemed happy just to follow us around. They held their tails aloft and panted lightly in the sunny, dusty yard. Harry was impressed by the place. He went up to a stack of immense beams, hundreds of them, each maybe sixteen inches square and fifty feet long. "These must have held up some huge building," Harry said.

Harry was happy in the beam yard. He patted the dogs and

the beams, and exclaimed at the salvaged stuff that lay heaped around the yard—an intact cupola, like a phone booth with its own tin dome, perhaps the peak of some public building; a round concrete relief of a baseball player, the bat across his shoulder. There was a heap of freeway guardrails, a pile of bedsprings, rolls of fencing and fire hoses—also a flagpole with a brass ball on top, lying askew on the pile. Mostly there were beams, though, in various sizes, thousands of them stacked stories high. In the canyons between them lay the sour smell of creosote and old beer, and occasionally a freight train shook the ground, the engine seeming to barrel right through the yard between the beam towers. Beyond the tracks was a corrugated building with a huge orange steel beam elevated over its yard, a big hoist, said Harry. On the orange beam, two words: IRON WORKS.

Harry found a length of bulldozer tread, lying flat on the ground like a plank. He tried to pick up one end of it, straining until the veins in his neck stood out. "This is incredibly heavy," he croaked.

The dogs wouldn't fetch. I threw a chunk of lumber for them, but they just continued to stare at me, as if I might eventually begin doing something significant, like feeding them. I pointed where I had thrown the stick, but they just looked at my finger. Hopeless, said Harry. He wondered if there was anything they would do. These dogs looked like they had rolled in grease, and then tried to clean up by rolling in mud. They just watched us. They followed us around very patiently, watching us. They were watch dogs, I joked.

After about an hour, the boss showed up. He pulled into the yard in his flatbed truck, waved and got out. He was a short, stout man, slightly resembling Buddy Hackett. The dogs abandoned us and ran up to him, and he took a couple of Dog Yummies out of his pocket and fed one to each dog. "What can I do you for?" he said.

"We need a beam. About ten feet long," Harry said.

"For a stock," I said.

"For a catapult," said Harry.

"It's an art project," I said.

The guy, who was almost as dusty as his dogs, listened to our explanation without remarking on it. He and his dogs just watched us as if we might eventually say something significant. "Just pick out a beam you like," he said at last.

"Where'd you get these beams?" Harry said. It wasn't a question that I would have asked. For me, the beams were the most prosaic matter imaginable, unremarkable objects, things one might see anywhere, every day, without ever seeing them. If you saw such a beam holding up a balcony, you'd look at the balcony, not the beam. The matter had always ended there for me. Ubiquitous was the word for beams, I thought, and like all ubiquitous things—the air, for instance—they seemed, despite their massive presence, to be almost invisible, warding off comprehension—mine, anyway—just by being there, thick and stupid.

But Harry and the beam boss really got into the beams. They spoke about them the way shell collectors would talk about shells. For the beam boss, the beams had personality and provenance. First of all they'd been alive. They'd been trees. He said one gray pile had been cut sixty years ago from old-growth redwood groves, the trees seven hundred years old and three hundred feet tall. They'd been seeded in the Sierra foothills in the Middle Ages, he said, cut in the boom days of the twenties, and built into his very own high school. The beam boss's brother ran the demolition end of the business, he added, and had gotten the contract to take down their old school last summer. "Tearing down your own high school," said Harry. "That sounds great."

One stack consisted of gigantic barrel staves from old wine vats in the Napa Valley. These were neatly stacked and bound, shims stuck between the planks to protect their surfaces. Harry noticed that the ends of these thick, perfectly straight-grained planks had been grooved, dadoed, he said, to fit the round top of the vat. They were black as ebony.

"That's clear heart you're looking at," said the beam boss.
"Was there a fire?" I asked.

"No, just time and tannin," he said. The surface of old redwood charred just from the oxygen in the air. Another pile of blackened boards was the remains of the last flume in Yosemite. They'd been cut from the first trees taken up there, and had been milled into planking for the flume, an elevated five-thousand-foot water channel used to power the generator that was for much of this century the only source of electricity in the park. Even after the valley was hooked up to the main grid, the flume stayed. They couldn't demolish it without further messing up the park—by putting in access roads to haul away the scrap. Besides, it was just too expensive, the beam boss told us.

"Then came the helicopter," he added dramatically. The boards we were looking at had been flown out of Yosemite by chopper. The flume had been chain-sawed into sections on the mountain, then flown out to the staging area, as the beam guy called it. He'd put in a bid on part of the load, and had sent a truck up into the hills to haul it down to the yard.

"And here they are," he said.

They didn't look like much to me. "Didn't they rot?" I said.

"Not these," he said. "You can just send these through the planer, and they're like brand new."

He took us to a big corrugated-metal shed at the back of the lot. When he opened the big door the light behind us fell into the space, illuminating the sawdust on the floor and a wide table-tool with a belt housing—probably the planer, I thought. Beside it was a steel cart with a single planed beam on it, pink as salmon in the light. Harry went in and put his hands on the pink beam, rubbing its clear grain.

"Let's get it," I said.

"You can't have this one," said the beam guy. "I've got a monk who takes these. He's building a temple out of them in the Santa Cruz mountains. I save him the best beams."

I went up to the perfect beam. It was sharp-edged and

smooth-faced and showed no knots. Both ends were still blackened, but the grain on the sides ran unbroken through all twelve feet of its length. It was a sliver of the heart of some huge tree, I thought. And as my eyes adjusted to the dark in the shed, I could see that one long wall was covered by a rack of wood, all candidates for the perfect beam. I asked about these, too, but the beam guy didn't seem anxious to part with them. Harry spoke to me sotto voce. "These are expensive," he said. "We're talking hundreds of dollars, Jim."

So we went back outside to find our piece of wood. Fir, not redwood, Harry said. The dogs rose and stretched at the edge of the sunlight as we came out. Harry and I climbed over and rooted around in a nearby beam pile, occasionally tipping a loose beam, until we saw one that seemed right. It was wedged under a whole stack of others, but seemed to be the right length, anyway. The beam boss didn't seem at all fazed by the tons of other beams that lay on top of ours. He went off with his dogs to get the forklift, as Harry and I peered into the stack, trying to get a closer look at the beam we had chosen.

The beam boss came back bumping over the dirt in his black-and-yellow Caterpillar forklift. He seemed very happy to be moving his beams around, though he'd left the dogs behind to keep them out of potential squashing range. The beam boss yelled orders to Harry and me from the caged cab. "You guys each grab an end and throw it out of there when I lift the pile," he said. Then he jammed and jiggled the fork into the stack at the level of our beam, raced the engine, and raised the entire quivering pile above our heads. The rear of the forklift somehow remained on the ground. Harry must have sensed my anxiety, as I hesitated to reach beneath the huge, tipping stack, because he shouted at me over the roar of the forklift.

"Just get the beam," he yelled.

We grabbed it, lifted it over the others in the pile in front of it, and dropped it on the ground near the front wheels of the forklift, then got out of there, as the beam guy lowered the stack, which flexed and rocked as it settled back down.

So we had our beam. It was covered with dirt and grease, and we smeared our hands lifting it. But Harry and I bore it between us on our shoulders, back through the lot, beam hunters heading home, followed by dogs. We put the beam into the bed of the pickup, but about four feet of it still stuck out the back, which Harry thought dangerous. So we anchored the back of the beam into the truck bed and tied the front above the roof of the cab, making the truck look like a missile launcher. The whole beam business had made me unaccountably happy, I thought.

And the beam guy proved to be a prince. Harry wondered aloud on the way home how he managed to stay in business—he'd spent an hour walking around the lot with us, discussing beams and life, and at the end only charged us sixteen dollars for the beam.

By the time we got the thing back to Harry's porch, our shoulders, arms, hands, even our faces bore beam grease and soot. I tried to wash, but I couldn't stay away from the beam. I looked through Harry's tools until I found a kind of double-handled blade, and began to scrape some of the grime and ash off of it. When Harry came out, he found me scraping.

"Cleaning the beam?" he said sarcastically.

"Yeah," I said. "This thing works perfectly." I held up the double-handled tool.

"That's what it's for," he said. "It's called a drawknife."

The drawknife carved off the blackened outer layer of the beam, and swirls of gold rose underneath. "Fir's gold," said Harry.

Susan came out to see our beam. "That looks great," she said.

"Hey," said Harry. "Let me do that."

"Just wait your turn," I said.

So I finished one side, and gave the blade to Harry. He rolled the beam over and started scraping.

"Gee Harry," I said, as he worked. "You're getting so aesthetic about this."

He didn't answer. He was too busy cleaning the beam. "Look at that," he said, when he'd finished.

CHAPTER 14

The Woods

My father taught me to play golf. He had some of his old clubs cut down for me—I was about fourteen at the time—and took me out to the club course. He instructed me on follow-through and showed me the interlocking grip. I hated it. I had no patience, and even at fourteen, I felt that there was something smug about the sport. We played in the wet heat of Virginia summer, the fairways manicured, shimmering, too green.

The only thing I liked was driving, stepping up to the tee with the biggest wood and socking the ball as far as I could into the complacent green distance. I could hit long and hard about once in ten tries, and then not straight, but I didn't care. Watching the ball slice over the trees and into the lake was my private satisfaction, scattering the cardigans around the next tee my secret joy. "Too strong, Jimbo," my father would admonish, as I watched the ball disappear in rage and glee.

If I could I'd talk my dad into taking me to the driving range so I could practice slugging. More often he'd insist on instructing me in the short irons, and we'd hit chip shots for hours until I would scream and cry to go home.

My father loved to instruct me at golf. It was a joy for him second only to golf itself. He instructed me at golf, whether I wanted to be instructed or not, for nearly twenty years. I'd be down in a sand trap, and he'd be standing above me, demonstrating with his own sand wedge, and instructing away from the lip of the hazard. "Head down, now. Widen your stance. That's an uphill lie, remember. Flex your knees." Finally I'd be so enraged that I'd swing too hard, spray sand everywhere, and club the ball over the green into the woods—either that or I'd miss the ball altogether, throw the stupid wedge down and stalk back to the clubhouse in tears, my father pulling both carts and pursuing me, telling me again what I did wrong. This was torture by golf, I thought.

None of this rage disturbed my father. He was pretty wooden, my dad. Cheery, cordial, great with strangers, but pretty stiff at home. But even being wooden has its virtues, patience chief among them. I could drive four shots into the lake, and his voice would resume behind me, telling me again that I should close the club face if I wanted to avoid that slice.

So I devoted much of my teenage life to making this man lose his patience and show his anger. It was my main work during that time, though I rarely succeeded. Occasionally, after I baited him for days, he would flare up, say God Damn It, and shortly return to reason.

Only once did I succeed in wringing violence from him. I had locked myself in my room and refused to come out. Behind the door I taunted and screamed at him. I kicked the door. To my amazement, he kicked back, not just once but several times. He kicked a hole in the door—his foot came right through the masonite. I was astonished, enraged, delighted. I attacked the door from my side, and together we took most of it out, battering the wood with our feet and hands until we could see each other through the hole. By then he was wooden again, telling me that this was all so senseless, that I should just try to calm down. I didn't calm down until the next day, by which time I had single-handedly installed a new door—a heavy solid-core one this time.

Golf was his real passion. He laid off only for the short Virginia winter, playing in the fall until it snowed, and in the spring at the first thaw. All summer he played—seventy-two holes a week, often—in the steaming heat. Saturdays he'd get up before dawn, meet the other guys in his foursome—Chuck, Marshall, Dick—and get in nine before nine. He was Bob. I'd wake up late and imagine Bob out there, taking out his driver, teeing up his Wilson.

On Sunday he made me go. I spent much of the day in the deep rough, which is like the rainforest in Virginia, looking through the jungly undergrowth for the godforsaken golf ball. If I was going to hit them wild, he said, I had to try to find them, and I'd end up in the bushes for hours, itching and sweating, while somewhere in the jungle behind me my father would stand and call out to me, instructing me even in how to find my ball.

"Jimbo," he would say, "comb the brush with your wedge."

When I was growing up, I imagined that my father had committed some awful crime as a teenager, and was spending the rest of his life keeping his murderous temper dutifully under control and scrupulously avoiding any mention of the past. This was a ludicrous explanation, I realized even then, though a comforting one. Anything was better than imagining that adult men simply turned to wood. I would not be like my father, I pledged.

He had grown up in Riverside, California, I knew that much. His father had run a sporting goods store; his mother was a dental hygienist. We went to Riverside only rarely, the town's name ironic, a real estate ploy. Riverside is in the desert. The river is a dry concrete trough, or at least it always was when I saw it. I never saw a photograph of my father as a child there. By the time he appears in a picture, he is about nineteen. He

is kneeling somewhere in a California desert landscape, posing
with his buddies next to the wheel and fuselage of an airplane,
a B-17 bomber, the famous Flying Fortress. He is wearing a
leather flying cap, and has goggles on his forehead. By then
he was already in training for the war.

He was a bombardier, and was stationed in England. He
was one of those Yanks, sent out in the night skies over Europe,
the planes often not making it back across the Channel, or
limping home over the white cliffs of Dover, pierced with flak.
Part of each mission he rode in the bomb bay, back in the
freezing belly of the plane. He had to arm the bombs by re-
moving a cotter pin from each one, and once they were armed
he peered through the bombsight in the aircraft's nose for the
abstract patterns of lights which meant towns below, and
dropped the bombs on command, pulling a lever and opening
the dark compartment to the wind.

Though he never said anything to me about the war, he
saved those cotter pins, and I found them years later in the
crawl space under the house, in a wooden chest stenciled with
his name and the abbreviation for lieutenant. There were
hundreds of the pins, each labelled with a tag that was tied to
the pin with string. On the label he wrote the name of the place
he bombed, the date, and sometimes a short note. One read
"Hamburg, July 28, 1943, Some Flak." Later I looked it up.
The survivors of the Hamburg raid had named the phenomena
they'd witnessed "carpet bombing." It started a firestorm. Said
one firefighter: "It was as though we were doing no more than
throwing a drop of water onto a hot stone." Twelve of 791
planes went down, I read. Operation Gomorrah, it was called.

Later I read Randall Jarrell's poem about the ball-turret
gunner: "I woke to black flak and the nightmare fighters. When
I died they washed me out of the turret with a hose." And I
knew that my father must have witnessed awful things, even
at his remove. But he never spoke about it. Once when I was
older I pressed him gently on the subject, and he said that the
English officers wore mustaches, which he'd never liked—said

jovially, as always. Jovially too, he once told me that he thought that the claims of a Nazi holocaust against the Jews were exaggerated. He did not want to see, I thought.

After the war, my father returned to California, resumed his education at Berkeley, met my mother, fathered me, and put the war behind him. In America, our cities intact, our economy booming, it was easy to do that. He had done his war work at high altitudes. And it had happened over there, which seemed like a long way off then. Most of all, we had won, and among the principal spoils of victory, it seemed, was the privilege of controlling the meaning of the war and, accordingly, of canceling its tragic claims on the future. Only for the losers was obeying orders no excuse. Europe had to live in history, but in America, in California, the war soon seemed ancient, almost legendary, as glamorous as *Victory at Sea*, as remote as Roman legions.

When I left home, I happily said good-bye to golf. I would never touch another golf club, I vowed. By then I had the term *bourgeois* to help me indict the sport, and with the Vietnam War in the background, golf seemed like artillery practice— chip shots like mortar fire, drives like guided missiles. When Coppola put surfing in Vietnam, I felt that golf would've been more appropriate—lines of golf carts proceding down the defoliated fairways of the delta.

I spent two years at the height of the war on a draft deferment as a college student, opposed to war in general, and to the Vietnam War in particular, as an intervention against a popular uprising. My father said my explanation was too simple, and asked that I not protest against the war in public. I might jeopardize his job, he said. I defied him, of course.

I was nonviolent, I thought. I would be a Conscientious Objector. And I began preparing a file, but as an occasional Unitarian—not a Quaker or a Jehovah's Witness—I stood almost no chance with my draft board, Virginia generals who could draft a Guernsey cow, we joked. Maybe I'd move to Toronto for the duration, I thought. But in the end I didn't have to.

One night the guys in my dormitory gathered in our TV lounge to watch a general pull numbers out of a bowl. It was the first draft lottery. The guys with low numbers would go. Afterward, a bunch of us went out to get drunk, about half in celebration. My birthday was picked as number 188, which seemed safe enough, so I went 1-A. After the lottery, 1-As remained vulnerable to the draft for a year, as the Pentagon called the numbers. By Christmas, number 182 had been called, but when New Year's Eve passed, the number hadn't reached mine, and I was clear. I had turned the corner on Vietnam.

On my visits home during college, my father would often ask me to play golf with him, but I would always beg off. "I'm only here for a week, Dad, and golf takes forever." So years went by, and I honored my pledge not to touch a golf club. But as my twenties proceeded, I began to soften both toward my father and toward golf. And my father had softened as well. He seemed a little less wooden. My youngest brother, I noticed, actually enjoyed playing golf with him. And by then I had decided that I wanted to spend some time with my father. So I guessed I'd play golf.

And I'd be instructed. Things hadn't changed that much. Whole decades had passed, but he was still reminding me to keep my head down. And I still enjoyed nailing the ball— whacking one over the green and over the pond beyond the green, while my father sucked in his breath, said, "Whoa ball! Whoa ball!" and finally barked, "Fore!" I was still too strong for my own good, my father said.

But I began to appreciate a shot that stayed in bounds. I found, to my surprise, that I could lay up a chip shot if I thought about it, and I didn't complain about being instructed. If he wanted a son to give golf lessons to, then I could be that son. He'd played golf forever, after all, and sometimes he'd even help, I found. And once I let him help me, his instruction seemed to become less evangelical, easier to take. Even so, I had the sinking feeling that this was going to be it for us, that golf was going to pretty much sum up what we could be together.

Maybe I was wrong, but I never found out. His death intervened, the year I was thirty. I called him long distance at the hospital, and we had our usual sort of conversation. He hadn't said anything that made me think it was serious— besides, he'd been on the golf course only a month before. So I just wished him luck with the operation. And later tried to remember what I had actually heard, the words he'd used, as I had spoken to him. He would have died soon, in any case, said the doctor.

Three or four years later, I took a junket to Fiji. I went with a group of men, all of them older than I was, hardened journalists being rewarded for their service by this lavish and embarrassing perk from their papers, the *Denver Post* and the Chicago *Sun Times*. In the airport lounge they told newspaper war stories. And we ended up in paradise. Girls in grass skirts brought us drinks by the pool. During the day we snorkeled among the beautiful fish. At night we wore flowers and straw hats, ate roast pork and mangoes. On our last night we all sang with the band.

There was a golf course at the resort, and our hosts were only too happy to provide us with clubs. I refrained until the last morning, the day we were to leave. Then I went out early, before the tropical sun was up, and played alone. For the first time ever, the game of golf was mine. The ball behaved flawlessly from the first drive, when it leapt from the club, flew the length of the fairway, and, collapsing off its trajectory, seemed to hover onto the apron of the green. Shot after shot followed, exactly. For once I could putt. Beyond the tropical rough, waves fell into the sand; on the greens, the ball clicked into cup after cup. I cursed my father and kept playing.

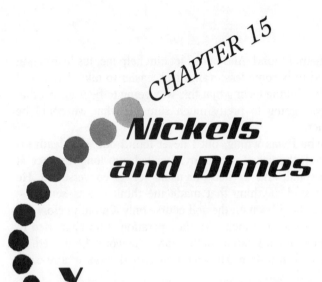

CHAPTER 15

Nickels
and Dimes

You can get the lag bolts," Harry said.

"Lug bolts," I said.

"Not lug bolts, lag bolts," said Harry. "Lag bolts, lag bolts— haven't you ever heard of lag bolts?"

Harry wanted me to go out and get some small stuff. The lag bolts, I learned, were long thick wood-screws with hex- agonal heads, which we would use to secure the wings to the beam. I had to get fourteen lag bolts and a drill bit—the bit to sink some holes in the beam to put the lag bolts in, Harry said. We needed a butterfly bit, he said, and I had to write down the size.

"How come I have to get all this stuff?" I said.

"Because you put the money in your account," said Harry.

At this point in the project, I still had confidence about the money. After all, we had five hundred whole dollars to build the catapult. But Harry and the small parts woke me from this financial stupor. Until we got to small parts I thought we might come out of the project with some of our materials budget left over. I dreamed Harry and Susan and Sara and I might even go out to dinner at Chez Panisse on it. And I thought we had good reason to feel assured. After all, the big stuff, probably ninety percent of the weight of the project, had only come to a hundred and sixty-five dollars, not counting the holes, of course, which had cost more than all the big stuff put together. The holes might've told me something, but they didn't. The lamb or the duck?—that's what I was wondering.

I called a hardware store in my neighborhood, but they didn't have any lag bolts as big as the ones we needed, and they suggested a store simply called Construction Device, which was south of Market near the Hall of Justice. On my way over there, I realized I'd never considered the number of small parts we were going to need. I counted them up in my head, just the ones for this first front-end assembly, and came to thirty-six: four springs, two spring clips, four bolts, four washers, four nuts, two I-beams, two plates, and fourteen lag bolts. There would probably be ten small parts for every large one, I thought.

Then I had the horrible insight, and dismissed it at once, that the cost of the small ones would outprice the big ones by nearly ten to one, as well. That couldn't be, I thought. The small parts were small, after all. Surely ten percent of the weight of the catapult couldn't compose ninety percent of its cost. And I hadn't even figured on the expense of buying the special tools we'd need, like this butterfly bit. I hated this kind of diminutive and mincing thought.

I found the hardware store, which turned out in fact to be right next to the Hall of Justice, across from the offices of several bail bondsmen. The sight reminded me of something else I had been trying to forget. In my briefcase was a copy of our permit to use the site in the Headlands. It had arrived in the mail two days before, and I didn't want to show it to Harry. It specified the date, time, and location of our firing, and in large letters in a box marked "Special Conditions" a sharp, authoritative hand had written, "WILL SHOOT 'MOCK' ROCKS." Harry would flip if he saw it. We hadn't even thought about mock rocks for weeks, while we had proceeded with the real rock-thrower. I had stuffed the document in my briefcase, but it hadn't gone away. I decided to try to forget about that, too.

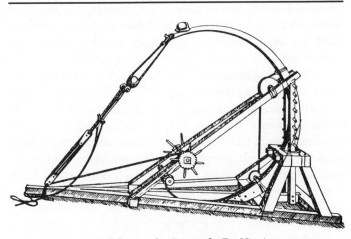

Rockthrower by Leonardo Da Vinci

From *The Crossbow*, by Sir Ralph Payne-Galway,
The Holland Press, Ltd., London, 1903

The clerk behind the counter at Construction Device brought me my fourteen shiny lag bolts, screws as big as link sausages. And he directed me to the drill bit aisle, where I found the bit we needed, hanging in its red plastic envelope among the ranks of other bits. It was a slender tool, not overly impressive, but it cost $16.95, more than our massive beam had cost. The lag bolts, thrown in a cardboard box and surprisingly heavy there, were two dollars apiece.

I got a bad feeling when I saw the register light up with the total: $47.87. I was doing that mental squirm one does, with the thought that there may not be enough in the bank to cover the check one is about to write. For the truth was that in the interim between getting the grant and paying for these parts, I had paid my rent, which cut into things. I figured I would repay the catapult fund when the checks finally arrived for some articles I'd written. Anyway, I squirmed in the bolt place, and I couldn't write the check. I put the lag bolts and drill bit on my credit card, praying I wasn't already over my limit, and got out of there.

On my way to Harry's, I drove across the Bay Bridge, its superstructure studded with thousands of bolt heads, I noticed. No wonder it required tolls forever. In the box in the backseat, the bolts clinked as the car thumped over the expansion joints in the bridge. At least the car was paid for, I thought, remembering for a second the misery of those car payments.

The catapult was going to cost us money. The five hundred bucks, that stunning largesse from the art center's board, the five hundred whole dollars was not going to make it. We were going to have to spend our own money on the catapult. Harry and I would probably spend at least a hundred dollars each, I figured.

After a while, trying to ease myself from this spasm of financial worry, I considered the matter platonically. These fiscal anxieties were the unavoidable birth pains of the catapult, which was entering the range of the real, and had to become, among other things, a pile of pennies representing labor and materials. I tried to observe the whole thing from that distance, as an example of abstract economics. But it was a hundred bucks. Or about as much as dinner at Chez Panisse, I realized with a pang.

At Harry's, I hauled the box with the lag bolts and the bit in it up the front steps and pounded on the door. Harry opened it. That day he was a pale gray, like a dove gray. The paint outlined a triangle of his usual skin color around his nose and mouth, where his respirator had been.

"Spray stuff gray today?" I said. He had painted himself in the process of painting shelves for a mountaineering store, he explained.

"Tomorrow I got to speckle them," he said. "To make them look more natural."

"So will you, I guess," I said.

We took the box to the back porch, where the beam and steel lay. Harry said he didn't really feel like working on the catapult at that moment. It was cold on the porch, and it was already getting dark. I said, "Just let me use the drill, then. You can just relax and talk to me while I do it." To my surprise, he agreed.

But of course he couldn't stay out of it. We marked the beam for the holes, to match the ones that Harry had already drilled in the steel mounts. Then we set up the drill, lay the beam on its side, and began boring holes in it, taking turns. But first Harry instructed me a little. We had to make the holes as vertical as we could, he told me. "Straight down on it, like branding a cow," he said, as if I had done lots of that. Solemnly, we drilled two holes.

As he was placing the drill for a third, I said, "Harry, that drill bit was kind of expensive."

Harry concentrated on the drill. "How much was it?" he said.

"With tax it was almost twenty bucks," I said.

"It's a tool," he said. "Tools are expensive." With that he pulled the trigger on the drill, and began boring into the beam. Then gold chips and ribbons of fir were pouring out of the hole where the drill spun.

"Maybe so," I said, raising my voice against the whine. "But if it keeps up like this," I said, "this project's going to cost us a fortune." Harry finished up the hole and pulled out the drill, which accelerated in the empty air, then wound down when he released the trigger.

"It's going to cost *you* a fortune," he said. And before I even had a chance to sputter, he added, "Speaking of money, you owe me a hundred and twenty dollars. I paid David for the springs, remember?"

"I thought we'd settle up at the end," I said.

"No way," he said. "Lookit, this whole oddball project was your idea, and I'm not laying out any money on it." Calming down, he added, "I painted myself gray today—do you know *why* I painted myself gray today?"

Behind us his big house loomed in the gathering darkness. A child would yell in there, if we waited a minute. I took over the drill and started on my own hole, berating him gently as I did. How could he say that? He was my partner. He was my friend. I had gotten us a grant to do this project. It was fun. He had to admit it was fun. And I thought we were going to share this experience. Harry with his dove gray head just looked impassively at me until I finished drilling the hole. He knew he had me.

"We're sharing this experience, believe me," he said, adding in a clipped, conclusory, and self-satisfied manner, "We're just not sharing the bill."

What could I do? I couldn't force him to pay his half—not with his appeals to house and family in reserve—so I tried as a last-ditch effort to appear saintly and generous, hoping he'd feel awful and renege. As he was drilling his next hole, I got out my checkbook and quietly wrote him a check. I told him I appreciated his help, and I gave it to him. But Harry just took the check, folded it in half, and said, "Thenk yo veddy much." I made a mental note to move the last of my savings into my checking account in the morning, and drilled my next hole with a vengeance.

But after a while, this thing we were working on began to draw me out of my funk. I had to concentrate on the drill and make good holes, and my mood began to lift as I did. Damn the small parts, I thought one last time. Damn the money. Then we finished up the drilling on one side of the beam, having stitched seven holes into the gold wood, and bolted one set of springs to that side. To screw in the lag bolts, we used a socket wrench with a breaker bar—an oversized handle like a tire iron. Even with such a big lever we had to work. The holes were tight, and the wood gathered its strength to resist as the bolts went in. On my first try, I couldn't get the bolt all the way down. I pulled as hard as I could on the breaker bar, and still a gap stood between the bolt head and the plate beneath it. I braced my feet against the beam and strained against the lever, grunting and cursing. I was not going to let Harry do this, I thought. I was not.

Harry let me sweat over it a while, then said, "Here," and jostled me off the bar and finished setting the bolt.

"See," he said. "You need me."

"So you think," I said.

We worked on the other bolts, and became absorbed. We were putting something together, and assembly had some basic suspense in it. As we put the wings on, we began to work faster and say less. With one wing on, the thing looked like a huge mattock. With two, it resembled an anchor, a pendulum, a cross. Whatever it was, though, it looked singular. It was a single and simple mechanism, still useless, but whole.

I hadn't counted the big beam as a part before, I realized. I had assumed that being so big it had to be something else, the background or something. But now it was a part with the others, and all of the parts together were one tight thing, a single symmetry, like the body with its limbs. I braced the thing with my foot as Harry sat on the stock and tried to pluck the bows. When he let go I could feel the big beam vibrate through my shoe.

I drove back across the Bay Bridge wondering what I had observed that night, and thought of the shavings that had come out of the holes, and how the beam had come to sit in a nest of them as we had rolled it over and installed it with almost everything else on the porch, composing our engine. No good as an observation, I thought—the shavings weren't even part of it. There was a universe of catapult parts, I thought, in which what might be useful to the project—the steel, the stones— existed. The wood shavings were castoffs, neither parts nor tools, and so they had ceased to exist in the catapult universe. They'd disappeared, not even part of something larger than our intentions at the time. A machine was the organizing principle among its parts, I observed, a proposition in which all the terms had to follow each other, a formula.

But that observation had no weight, either, I knew. Because as I thought about it, I had to confess that the catapult's meaning hadn't really come up that evening. That we could afford it, that we could make it work, those were our concerns. That thought gave me the sinking feeling, as I drove home beneath the Bay Bridge's towers and cables, that I wasn't outside the catapult anymore, seeing it as an object in the world, but inside it somehow, assuming the world from its viewpoint. As if we were its eyes, its mind, on the lookout for the best way to make it manifest. And that was no observation at all—just a kind of mechanical vision, fascinating and persistent, against which my need to observe was an eddy to the main current. By the end, in fact, I would add up our expenses and be astonished, chagrined but secretly proud that the expenses had gotten so far out of hand, that I had spent six or seven hundred dollars of my own money on it.

In the morning I called Lennie, the accountant who does my taxes. "Listen, Lennie," I said, "I've got a bunch of expenses for a project I'm working on, and I'm wondering if there's any chance they might be tax-deductible."

"No problem," said Lennie. "What are they?"

"They're hardware, mostly," I said.

"It's a computer project?" asked Lennie.

"No," I said. "This is real hardware. Old-fashioned nuts and bolts. I'm building a big catapult with a friend."

"One of those knights-in-armor things," he said, flatly. Lennie had heard everything, I thought.

"Right," I said. "I think I'm going to write about it."

"You're writing about it?" said Lennie. "No problem."

"No problem?" I asked.

"No problem," he said. "You're a writer. Writing's your business. If you're writing about this knights-in-armor thing, then you can write it off as a business expense. Besides, building a weapon the feds will understand."

CHAPTER 16

The Destruction of the Second Temple

In 63 B.C., the Roman Pompey reclaimed Jerusalem for the empire with a minor siege, getting help inside the walls from a branch of the royal family. Nevertheless, Pompey had to complete his conquest, and so he went to the temple of the Jews and commanded an entrance to its innermost chamber, the location of the Holy of Holies. The act meant death for an ordinary Gentile. Even among the Jews no one but the high priest might enter into that place, and he only on the Day of Atonement. But Pompey was not ordinary—he was deified, after all, in Rome, and he had besieged the town with catapults, besides. So might made right, and he gained his entrance and lived to make a report.

Inside he saw nothing, and was amazed. The temple of the Jews, the largest and finest shrine in the world, had an empty space at its center. An idol he might have understood—he was one himself—but within the sanctuary, Pompey beheld nothing and made nothing of it, a mistake. He had not, apparently, read the Second Commandment, the one that prohibited making likenesses, images "of anything in heaven, or on earth beneath, or in the waters under the earth." Inside the sanctuary there could be no image, and what was left was emptiness, just the place, the stone and the space above it. What good was conquest, then? Pompey might have asked himself. Nonetheless, the site of the Holy of Holies was the place above all to be defended, hence to be besieged. And in the year 70 A.D. its destruction by siege would change the course of humanity.

The place of the Holy of Holies is upon Mount Moriah, the Temple Mount in Jerusalem. By tradition it is the rock upon which Abraham sought to kill his son Isaac, where God set aside that sacrifice. It was a place stained by the blood of human sacrifice before that, the stone holy to Moloch, the old god who demanded the killing of sons. After Abraham's hand was stayed only animals were offered at the altar.

King Solomon built his temple to house the site, the chief work of the reign of peace and art that followed his father David's time of war and conquest. But Solomon's temple burned, besieged by Nebuchadnezzar and his Babylonian army, who set fire to tapestries depicting the stars, and sent the Jewish king Zedekiah to his fate—his sons killed before him, his eyes gouged out, he was made into a traveling exhibit of the might of Babylon. Displaced, the Jewish survivors fled or fell into slavery, and with the perfect cube of the sanctuary reduced to rubble and ash, the place of the Holy of Holies went unhoused for centuries. From Babylon, the exiled Jews vowed to remember and return. "If I forget thee, O Jerusalem," went the pledge, "may my right hand forget its cunning."

A half millennium later, the Jews had indeed returned to Zion from their Babylonian exile, and had built a second temple on the place of the Holy of Holies. By then, the ruling Jewish family, the Hasmons, had risen in a revolution against yet another wave of empire, this one from Greece and Alexander, Aristotle's student, who had quashed the East with his new catapult. But Alexander's power, too, had waned, and the

Hasmons had risen against the Greeks, led by Judah, the best fighter among his brothers and so called the Hammer. A century later his descendants all felt royal, and were fighting among themselves. Pompey and his empire intervened, and seemed to settle the argument.

But arguments in Israel don't seem to settle. Pompey had poisoned his authority over the Jewish masses by defiling the sanctuary, for one thing. The ruling family he left behind in Judea were collaborators, many thought. After that, loyalty was fatal, and with the insiders thus compromised, enterprising outsiders did well. One of them was Antipater, a chief minister from the province of Idumaea, a province brutally conquered by the Hasmonaean dynasty. As the Hasmons fought among themselves, the outsider Antipater was able to ally the country with Rome—with Pompey—and consolidate his own power in the new order. When he did, he appointed his son Herod governor of Galilee, and assigned him to impose order on the guerrillas—or the bandits, as you like—there in the wilds. Thus began Herod's murderous career. Ultimately, with Rome's help, he would be king.

There can be little doubt about Herod's brilliance as a politician. In his heart of hearts, he was probably still an Idumaean, the son of a conquered provincial and a forced convert, and this was the right background for his principal task— serving Rome while mollifying the Jews, especially by honoring their religion. If he failed to serve Rome, he would be ousted; if he alienated the Jews, the priests might lead them in rebellion. He responded by the time-honored method of funding enormous, labor-intensive public works projects. In the provinces, he built Caesarea, a new town, an administrative center from scratch, like Brasilia. He rebuilt the Temple of Apollo at Rhodes, pleasing the country's bourgeois Greeks and appealing to the Romans' Greek pedigree. And for the Jews, he enlarged and glorified the temple in Jerusalem, until it eclipsed even the glory of Solomon's. He had the sanctuary area screened from view, and took the old building down. Then he built massively up, stacking enormous crisp-edged blocks of local stone into walls hundreds of feet high.

In Herod's time, the multitude roared there, herding their animals up to the temple for sacrifice. Some bought doves for killing when they changed their money into holy tokens in the plaza below. Then, among the bleating flocks, they crowded the broad staircase running the length of the temple's south side. At the top of the stairs they went in beneath a high, thick double arch, had their beasts' throats cut, and exited renewed beneath the triple arch farther down the same long, southern-facing wall. Atop it, above the arches, was a long classical court, the chambers of the Sanhedrin, the supreme council. And behind the council chambers, the plaza on the top of the Mount opened, its bounding colonnade enclosing the sanctuary itself. The plaza bustled with administrators, who passed briskly down the vaulted corridors between hundreds of thick Corinthian pillars.

Public signs prohibited Gentiles in both Latin and Greek, but foreigners might make sacrificial offerings, to placate the local god. The consul Marcus Agrippa offered his hundred-beast hecatomb, the blood sluiced away by streams of water from the temple's elaborate reservoirs. To further honor and welcome such Roman devotions, Herod ordered a golden statue of an eagle placed over the temple's main gate, where it remained an abomination to the devout until at last Herod himself fell mortally ill. Then a group of Torah students pulled down Rome's golden bird, smashing it. From his deathbed, Herod ordered that the students be rounded up, taken to the stadium at Jericho, and burned alive. Then he himself died, in transit, as he was being borne on a litter toward his spa in the desert.

Whatever may be said against Herod, his vicious and titanic appetite for power tended to unify the country. He had counterbalanced the empire and the Jews, and without him the two began to collide. Rome quartered the state and cracked down. At the temple, the high priests outlawed sacrifices by foreigners. The Jewish factions split violently. A few took knives beneath their cloaks into the marketplace, and murdered the friends of Rome. Rock-throwing bouts broke out, and the empire bristled and made martyrs—Jesus among them.

The time felt like the End of Days, impending doom instructing prophecy everywhere. Gazing upon Herod's temple, Jesus foretells its destruction, saying that not one of its stones shall remain upon another. His crucifixion was typical of the Roman means of execution, a technical display for the locals— the Jews preferred public stoning. Mounting the knoll of Golgotha beneath His cross, He bids the daughters of Jerusalem, "Weep not for me, but for yourselves and your children."

But it would be another thirty-three years before the chaotic situation in Jerusalem would finally break into the war that would consume Herod's temple. By then moderate opinion had become wholly untenable in the face of steadily worsening insults from Rome and its Greek proxies in Palestine. In Herod's new town of Caesarea, the Greeks built a factory cheek-to-jowl with a synagogue, then jailed the Jews who complained, and finally turned their celebration into arson and theft, a pogrom, Jewish homes sacked as the Roman garrison sat on their hands. Refugees jammed Jerusalem, and on the next feast day, someone in the crowd yelled to a visiting Roman dignitary, imploring him to liberate their country from its current corrupt and incompetent Roman governor, Florus. When the governor was told, he was incensed. Florus demanded that Jewish leaders surrender anyone who had yelled, and when no one came forward, he ordered his soldiers to kill anyone they found in the marketplace. Thousands died. And then, despite rational appeals and accurate estimates of what would happen next, the Jews rebelled in numbers large enough to matter, attacking the garrison in Jerusalem, slaughtering the soldiers and reclaiming their city.

Alarmed, the regional command of the empire mustered a

quick response. But Cestus Gallius, the Roman legate in Syria, underestimated the strength of the rebels holding Jerusalem. He marched upon the city with twenty thousand troops and a train of anti-Jewish rabble, and had his force cut to pieces. They fought five days beneath the walls, and finally fled under intense counterattack, down the switchbacks of the mountain path, jettisoning their baggage as they ran. Those who reached the bottom first found themselves trapped by a rebel force below. So the Romans spent a long day on the slope, pelted with stones and pierced with arrows until night fell and a few could escape.

The chain of ever-enlarging retaliations then reached the emperor in Rome. Nero was having his own problems and didn't need this Jewish victory in Palestine to appear to manifest the weakness of his rule. So the Jews would bear the full brunt of the legions. Vespasian himself, soon to be emperor, was given charge, and he placed his son Titus second in command. Together they took back Palestine in the Roman manner, inexorably, ponderously, yard by yard, leaving finished roads in their wake. Vespasian invaded at the head of a huge and technologically sophisticated force of arms, four full legions and hundreds of catapults and other engines of war, and in three years, he secured first the coast, and then the countryside. At that point he was proclaimed emperor, so he left his son in command and returned to Rome. And soon after, the Roman forces under Titus stood arrayed around the captive city of Jerusalem.

One may observe the depth of the tragedy to follow by the absence of honest record about it. Almost nothing from the Jews, who were afterward engaged in the anxious work of survival in alien lands. And from the Romans, very little. For one thing, these events were provincial—a volume on the outlands of Palestine might make kindling along the way. And for another, the field was preempted by a single, loud, partisan witness. The only so-called history of the destruction of the temple comes from a traitor, a dog, Josephus. His purpose was to please Rome. The real Jerusalem wavers in this lens.

Josephus was a man who would die in bed—and in a nice bed, at that, in his villa near Rome. A Jew, he wrote for his Roman patrons his version of the history of his crushed people, after he had aided in their crushing. He had begun as a leader among the Jews. In that brief, nervous interval—Rome tossed out, the Jews suddenly in their own kingdom again—Josephus took command of the defense of Galilee, where he fortified the walls and watched for the legions, then fled before them and finally surrendered. After that he was Vespasian's dog, flattering the general from the first. You look like an emperor, Josephus says he told him. A man after my own heart, thought the emperor-to-be.

At the siege of Jerusalem, Josephus was sent out ahead of the legions to talk his countrymen into surrendering. From beneath the walls, he beseeched the defenders, trying to convince them of the invincible might of Rome, the approaching juggernaut of catapults. "You stubborn fools," he says that he said. "Throw down your weapons." Hoots of derision from the battlements. "Take pity on your birthplace," he shouts. "Turn around and gaze upon the beauty you are betraying." From the towers, curses and showers of stones.

After that, Roman might was quick in arriving. Prince Titus and his vanguards converged from three directions at once, appearing on the ridges above the city. They began heavy construction immediately, improving their roads and building three walled siege encampments. The prince took his command from Mount Scopus, east and above the city, across the Kidron Valley with the Garden of Gethsemane in its depths, its olive trees green in the wet weather. From Mount Scopus, Titus might gaze down onto the colonnaded plateau of the Temple Mount itself. Atop the mount, the gold-and-white sanctuary, like a jewel box at that distance, stood in the temple's plaza, as bright, wrote Josephus, as the sun in the morning.

Much sheer work followed. Only miracles are achieved without labor, notes unctuous Josephus, as he describes the Romans' work of preparing their siege. The tense calm lasted for three weeks, a kind of peace broken by bowmen picking sentries off the wall and cavalry hunting down and crucifying any Jews who ventured outside the city in search of food in the valleys. Nobody was going anywhere; nor would help come from the hills. The walls rang with the sound of hammers, as the Roman engineers built their platforms and siege weapons and put up their own wall surrounding the city's, a bulwark four and a half miles long with thirteen towers, boasts Josephus for Rome, constructed in three and a half days.

When the Romans were ready, they moved their platforms—great roofed racks lined with catapults—into position beside the city's walls. From the platforms, the besiegers could fire down upon the defenders on the walls, hoping to drive them off the battlements, so that battering rams might be employed against the stone below. The Romans readied their biggest ram, which they called Victor.

The Jewish defense was furious. Some dug secret tunnels beneath the walls of the Citadel, then raced out into what had been the city's vegetable gardens—now enemy territory—bearing flaming brands, where they attempted to light the Roman platforms on fire, at the same time grappling and knife-fighting with the Roman guards. They actually succeeded at first, burning one of the platforms, astonishing the Romans, who responded by forming human barricades around the legs of their platforms. But the zealous defenders were undeterred, and the martyrs with their torches continued to leap into the midst of these close-packed brigades.

Meanwhile, their roof of shields taking arrows like hard rain, the Roman shock troops worked with crowbars and rams to undermine the wall. At the base of the Antonia, the city's lower fortress, troops managed to work loose four huge stones, and that night in the rain the whole wall collapsed. Inside, the Jews worked feverishly, putting up an inner wall, which they managed to defend with heavy losses the next day. The first Roman soldier up the wall tripped at the top and took a dozen arrows, says Josephus. After midnight on the night of the second day, the Romans tricked the Jews into sounding a full-scale alarm against the feint of a few scouts. Then Titus and the troops rushed the off-balance defenders. In they went, the crowds of combatants killing each other in the stone corridors inside the fortress, where they fought until the next afternoon, when the Romans had the Antonia secured.

Within a week they razed that fortress, laid it flat, and built a wide road through its ruins right up to the wall of the Temple Mount itself. Then the Romans brought up Victor, and pounded for six days against the huge stones of Herod's wall, without result. Finally they set fire to the main gates, melting the silver off them, and rushed the breach. Titus, conscious of the symbolic power of the sanctuary, closed in upon it with a few crack troops. When these had advanced across the plaza to within throwing range, one soldier climbed on the back of another, and hurled a flaming bough through the sanctuary's Golden Window. The flame took, perhaps on the tapestries inside, and some of the Jews ran to douse the blaze, forgetting the fight. The Romans overpowered the others and rushed into the burning sanctuary. More blades at close quarters: the dead fell around the altar, their blood slicking the marble steps.

Victorious Titus ordered the fire put out, and pulled down the curtain screening the place of the Holy of Holies. He took this screen back to Rome as a trophy of war—the ultimate prize was nothing, remember, and so could not be taken. A month of street fighting remained for the Romans before they could find no more resistance among the people of Jerusalem. Before he left, Titus ordered the city razed. He exempted only a small barracks from the destruction, that and the three towers of the upper Citadel, one of them named for Herod's wife, Mariamme. The towers might serve two functions, after the fall—to survey the ruins and to remind the Jews of the glory their city had been. So completely leveled was everything else, says Josephus, that no visitor afterward would believe that people had ever lived there.

Titus decorated his soldiers and left. In chains, the captive Jews became a great dying caravan of slaves and meat for the arena, following Titus's slow train home. Then, in a ruined Jerusalem, a few survivors finally emerged from their hiding places, tunnels mostly, to find ashes and corpses where glory had been. In Rome, the Senate ordered a Captive Judea coin minted, and a triumphal arch erected to honor Titus, on it in elaborate relief the images of a catapult and a menorah looted from the temple. Titus briefly succeeded his father as emperor, adored officially as "The Delight of All Mankind." Josephus emigrated to Rome, where he began his memoirs and added Flavius to his name in honor of Vespasian's dynasty, known to history as the Flavian. And the surviving Jews dispersed, stateless, into Syria and beyond, beginning their modern wandering, their diaspora. With the altars of the temple broken and lost, they made their sacrifices in prayer, and felt with new force the old grief of the prophets.

CHAPTER 17

The Comealong

Harry had been looking at a picture of a crossbow for inspiration when I got over there the next afternoon. He'd brought out his big, spine-split archery book and had laid it open on the stock of the catapult. The illustration he was poring over, from a medieval woodcut, was the top view of the crossbow. Harry thumped at it with his finger, and looked up at me.

"That's what we need," he said. "It's called a cranequin."

It took me a moment to figure out what he was pointing at, not at the whole picture, which was hard enough to see, but at a detail among the rough dark marks of the woodcut.

"You mean that crank," I said.

"Not just the crank," he said. "This whole thing." The crank, I saw, sat on a round housing, from which both ends of a toothed bar protruded. The bar was a rack gear, Harry explained. You turned the crank, and a gear inside the housing climbed the bar, tooth by tooth, drawing back the bow in the process. When the bow was fully drawn, it could be locked in place, and the cranequin taken off.

"Imagine them cranking it like crazy in battle," Harry said. "Arrows zipping past you. You fire the bow, put in an arrow, put on the cranequin, crank up the cranequin, take off the cranequin, fire the bow." He acted it out as he did it, like an organ-grinder in a state of mortal terror.

"Do we have to make this thing?" I said. "Or can we find a cranequin outlet?"

"We can use a winch," Harry said. "And a big one—look at this," he said.

Harry put the book aside, straddled the stock, and spread his arms to reach the tips of the bow. Then he pulled back with all his might, flexing the bows back a couple of inches. "This is going to be really, really strong," he said.

I tried to pull the springs back as well, and then we tried it together, sitting on the porch with the stock between us, each of us straining to pull back just one wing. The two of us did only a little better than Harry had. We had underestimated the strength of this bow, Harry said.

"How do we know how strong it is?" I asked, hoping this wasn't bad and expensive news. A jar with a stuck lid could be stronger than a person was, after all. But I could see how far these springs might flex, and how little we moved them, and this gauge allowed me to imagine just how many more times stronger the springs were than we were, and just how much power we might put behind the stones, given enough pull. Right away, I wanted to be given enough pull. I thought we might find a used boat-winch, but Harry said we could just get a comealong, a levered hand-winch.

"Let's go get one," I said.

First we had to figure out a way to anchor it, said Harry. "Where are we going to put all that force?" he asked me. The comealong could only be as strong as its anchor. And only our beam was strong enough to provide that anchorage.

Harry answered his own question. Looking at the woodcut of the cranequin, he had noticed how it had been anchored to the crossbow stock. What Harry saw in pictures always amazed me. Here was this blotty old woodcut, not clear in the least, and he read it like a dissertation. They'd stuck a bolt through the end of the stock, he said, to prevent the loop of cord attached to the cranequin from riding up the stock. We could adapt the idea by putting a couple of lag bolts into the base of our stock, and anchoring the comealong there. When we cocked the bow, the comealong sling would hang up on these bolts, its loop taut across the bottom of the beam. "That way it'll have to pull through the whole stock to come loose," Harry said. "It'll just bite harder and harder into itself."

After he explained it, I thought I got this part. Harry wanted to go on and talk to me about his thoughts on the bowstring, but I stopped him. I suggested that we were getting beyond ourselves. "Let's just get this comealong, and whatever we need to anchor it," I said. "We can concentrate on the rest tomorrow." I wanted to take one precise step at a time, while I could hold everything—barely—in mind. For the first time in the project, I was putting the brakes on Harry. He agreed, but as we got into the car, I could still feel him thinking, his mind clicking from this step on to the next and the next. I remembered his son Isaac's warning: "You better watch out— Papa will just take over." Now I felt that with some effort, I'd started a boulder down a hill. It was finally falling with its own weight, and gathering speed.

Feeney Wire Rope and Rigging was like a jewelry store compared to the other places we'd been. By this time I was starting to appreciate shops, and Feeney's was a beaut. Outside in the yard were stacks of huge spools, wound with ropes and cables thick enough to tie off an ocean liner. A showroom fronted the workshop and warehouse, five brass portholes set into the wall above the door.

Inside the showroom, the merchandise gleamed. This stuff was for yachts, I thought: nylon slings in pink, blue, and green, racks of brass cleats and clevis pins, spools of braided rope, a display of block and tackle that ran from clever little hand-sized models to whoppers as big as hams, pegboards of cables and chains and shackles. The place even had those hoop-

Fortress besieged

Reprinted in *The Crossbow*, by Sir Ralph Payne-Galway,
The Holland Press, Ltd., London, 1903

legged nautical chairs at the counter, and posters on the wall offering rewards for information about stolen boats.

Harry was picking up swivels and turnbuckles, exclaiming over the stainless steel aircraft cable, hefting the anchors. In front of one display he stopped. "Comealongs," he said. And there they were, about eight sizes of them. Each had the same configuration, a thick steel rectangle with heavy hooks at both ends, and a spool of cable or chain in the middle. A lever, hinged to the frame, turned the spool. One of the hooks, I saw, was attached to the end of the chain.

"Look," said Harry, grabbing one. "You just release the lock, let out the chain, grab what you want with this hook, and winch it in with the lever."

All but one of the comealongs had names like Jet, Ratcliffe, Little Mule and Lug-All. The unnamed one on the end was slighter than the rest and had a special tag that read, "Imported Comealong—$15.00." I looked at the price tags on the others. They were all more than a hundred dollars. We asked the guy behind the counter about it. "We sell a lot of those," he said noncommittally. "Sometimes you'll get into a job where you just have to trash a comealong. Maybe it gets hung up somewhere that you can't get at it. With that one, you can just leave it behind and not feel too bad. Throw it away when the job's over."

This guy's name was Rick. He was one of the partners at the place, and proved to be a real find. He had a round jaw and a big mustache, and he talked in big, informative, friendly paragraphs, like he was getting paid by the hour, Harry said. But what he said about the throwaway comealong wasn't actually too encouraging. What would it lift? asked Harry. "A ton," said Rick, "dead lift." Would that be enough? I asked Harry. Harry just shrugged. I was wondering—not for the first or last time—how much a truck weighs.

"What do you plan to do with it?" asked Rick. I told him we were pulling back a couple of sets of truck leaf springs. Why? Rick asked. We're throwing some rocks with them, I said. How big are the rocks? asked Rick.

"Big as grapefruit," I said.

"That sounds cool," said Rick.

It was a nice moment. Rick got into the project. He thought it was slightly loony, but cool, and he talked to us about it for a long time. "Jeez, a catapult," he said, shaking his head. He told us the cheapo comealong would probably do the job. We could probably pull two tons with it. Dead lift was another story, though, he said.

Rick left us alone to figure it out. I was all for getting the thing. I was sure we weren't dead-lifting. "Harry," I said, "two tons is four thousand pounds. We can't be putting that much on those springs." We took the comealong off the rack and messed with it. It was a cheapo, all right, the steel just thin plate, folded over to give it some added strength. The spot welding looked sloppy, too, Harry said. Still, we decided it would do.

We found Rick again and asked him about a loop of cable for our anchor, and he said he could make a sling for us while we waited. He went back in to the workshop, and Harry and I decided to get a length of heavy chain and some releasable links, like mountaineering carbiners. We'd use the chain temporarily to cock the bow, until we could figure out the details of the actual bowstring. When Rick came back, he brought us our sling—a piece of heavy cable with its ends clamped together by a steel sleeve—and I paid him for the stuff. It was another fifty bucks, but it was blood under the bridge at that point.

We stopped at a hardware store on the way back and got more lag bolts, and by the time we got back to the porch, it was dark. The big device lay there like the horse for whom Harry and I had bought this new bridle. By then our mental image of the needed work was very sharp, and we set to it without talking much, efficiently drilling holes in the base of the beam, putting in the new lag bolts, fitting the links first in the chain and then in the holes in the ends of the springs. Then we put our sling around the beam—just as the cranequin woodcut had shown—let out the cable from the comealong, and locked its hooks to the chain and sling. Then we were ready to cock the bow.

* * *

Harry took the lever of the comealong and clicked it back several times, until the cable and chain went taut. He clicked it a few more times until the bows began to bend.

"I'm not sure how far I want to cock this thing tonight," he said. "For one thing, I'm not sure how we're going to release it."

That had never occurred to me. "Don't you just unlock the comealong?" I said.

"It's going to pull out of there really hard," said Harry. "And I sure don't want to have my fingers in the gear when it does. Besides, we've never put any real tension on these springs before—maybe they're flawed. Maybe they'll break. Maybe the chain will break."

I remembered David's story of the slashed Asian woman. I had broken steel guitar strings, tuning them. Once the sharp end of a string had slashed my eyebrow open. I looked at the tense cable in front of us. "Let's pull it back a little bit more, anyway," I said.

Harry levered the comealong back four or five more clicks. By then the springs were flexed back about six inches. I touched the chain. It was rigid, as if it had been welded there. "Just a few more clicks," I said.

His face a little squinched, Harry clicked the lever back slowly, three, four, five times. That was it, though. We stood up to admire the thing. It looked good, solid and tense like that, and for a while we just admired it. Then Harry began gingerly to examine the springs. He didn't want to get in front of them, but he leaned over and looked at the place at which they were bolted to their mounts. "Damn," he said. "Look at that. They're torquing them."

"What?" I said.

"It's bending the mounts," he said. "It's bowing the I-beams back."

I noticed what he was talking about. The I-beams looked awful, suddenly, twisted and strained across their main brace. They were way off the perpendicular, said Harry. We'd have to fix that, somehow. The I-beam steel wasn't springy like the wings, and would absorb a lot of the power of the bow. "Let's get the stress off of them," I said, "before they break or something."

"They're not going to break," said disgusted Harry.

But we had to get the force off them, and Harry had no choice but to put his fingers down into the gears of the come-along to release it, one click at a time. The comealong had a sort of trigger, a spring-loaded pall, Harry called it, a tooth that lodged in the ratchets and kept the spool from unwinding. Under the strain of the cocked bow, that pall was stuck tight, and Harry had to work at dislodging it, at the same time assuming the stress on the spool with the lever. When he did manage to release the pall, he backed the lever gently down one notch, and reengaged it. He didn't want to unjam the whole thing at once, he said. First of all, he had his fingers in there. "Besides," he said, "dry firing is bad for a bow."

Once his arrow had fallen out of his bowstring just as he was letting it go, he said. Without the weight of the arrow to push, the bowstring oscillated wildly, snapping back in his face and breaking his glasses. So he daintily and nervously backed the comealong, his fingers in the gears. "You got me into this," he said.

After he had clicked it down a few times I couldn't take it any longer, and I said, "Just release the bow, Harry. It's mostly down anyway." He agreed, seeming relieved to get his fingers out of there. He got a screwdriver and began prying at the pall.

The comealong snapped, and there was a burst of release as the springs fired and a loud crack as the jolt threw the chain into the spring mounts. We jumped back.

"God, that was quick," I said.

"Whoa," said Harry. "We've got something here." He crawled over the mechanism, inspecting for damage. "We've really got something here," he said, holding up a dangling cable. "I think we stripped the comealong."

The force of the release had ripped the cable off the spool.

But Harry was happy. "You know what? We don't need to fine tune this thing at all. We don't need the pulleys. All we have to do is load this baby up and fire it. These are monster springs we've got here."

We jumped around and whooped. I was glad to be free of the task of making the eccentric pulleys. It was a technicality I'd been dreading, a flourish that even Harry later admitted would have added only ten percent to the power of the bow—and then only if we had accomplished it perfectly. But from this first comealong-stripping firing, Harry was sure we had all the force we needed. This thing was going to throw some rocks, he said.

The cheap comealong never wound correctly afterward. We had apparently exceeded its two-ton pulling limit. This was no jar with a stuck lid, I thought, though the force we had marshaled still seemed hard for me to imagine. Maybe all things stronger than a person take on a kind of conceptual equivalency, I thought, the weight of the whole earth equal to a stone one ounce heavier than you could lift. For that matter, the top of the sky could be just ten feet overhead, or eighteen miles up—it hardly matters if you don't have an airplane, and it's a comfort to imagine it just out of reach, as it is to imagine the earth to be six thousand years old. The telescope was a sort of lever, I mused, a kind of comealong.

After that, Harry and I managed with our stripped comealong for a while, though we had to take pains to reconnect and rewind it each time. When the day came to fire the catapult, we would rent a heavy-duty comealong for the shoot, and chuck the cheapo. Throwing it out, I remembered that Rick at Feeney's had recommended it by saying that you could throw it away if you had to. And you'll have to, he might have added, I thought.

CHAPTER 18

The Trigger Finger

Surfers talk about waves getting critical. If you try to get into a wave too early, its face is too gradual to push the board. If you get into it too late, the wave breaks on top of you. A wave is critical for an instant. Being there in that instant is the whole key to surfing, because if you stay where it's critical, you can ride.

When it came down to the last stages of building the catapult, it was clear we were riding. There was a point at which the catapult had gotten critical, too, I thought, after which everything followed. I thought of the desire to throw big rocks and the physical impossibility of doing so bodily as poles, with the catapult standing as a half-step between. Within that half-step was the half-step of the springs, and within that the engineering of the means of engaging them. Seeing the power of the bows harnessed was the critical moment, I decided, the moment it would be done. After that, our half-steps had proliferated, until at the end they were many and tiny. But the more half steps we took, I saw, the more fully the catapult would express our desire.

Harry had gussets on the brain. The sight of those I-beams bending under the force of the springs upset him. Power loss, he said, that's what that is. Those I-beams wouldn't spring back as readily—they were putty, relative to the spring steel. Harry fretted for a while and devised a pair of braces, which he called gussets, right-angle triangles of steel that we could weld into the right angles between the I-beams and the steel plate on the beam. Harry made two cardboard templates for the gussets and asked me to take them to David's. I said I would, but it was hard for me to consider something called a gusset very seriously. Just take them to David's, he said.

Then he raced ahead into the flurry of implications. Since there was no need for the pulleys, we could design a relatively simple bowstring; since we had proven that we could cock the bow, we could go ahead and make a trigger to secure and release the power we had harnessed. Approaching its completion, the machine gave us a ride, speaking to us unambiguously, as if broadcasting clear conceptions of what needed to be done. I didn't argue with Harry—I did things. If we saw shortcuts, we took them without debate.

I was buying hardware at a tremendous rate. I bought things that I couldn't even identify later from the bills. Among my purchases from Feeney Wire Rope were some seven-dollar items, identified on the receipts only as four number 02068s, and several eight-dollar things, number 00308s. Slips of paper bearing these numbers littered my desk. Writing among them, I had the giddy sensation of surfacing in scuba gear, rising within a column of air bubbles that grew denser and brighter as I came up.

We had become regulars at Feeney Wire Rope, having been there three or four times, each time getting to know Rick better as we talked over the project. One Saturday morning we went over there with Isaac, Harry's ten-year-old. Under the guise of giving Isaac a tour of the shop, Rick took us through the work area, narrating as he went. The place was immaculate, festooned with coils, the workbenches like displays under bright lamps.

Rick liked the idea of participating in an art project. He thought of himself as something of an artist, and showed us some of the things he had built for the shop. He'd made and painted a colorful hand-cranked device for spooling cable, a wheel radiating arms. Rick had constructed it out of spare parts, and the device looked like a Duchamp sculpture, though it turned smoothly on its gears and worked well. He spooled some cable for us with it, then gave the crank to Isaac, who turned the handle for a while, laying the strands of cable side by side inside the arms in neat coils, as we three men watched him.

We were there to check on Rick's progress. We had hired him to weave our bowstring out of cable—it was beyond us, we confessed—and he was taking his time. Rick had major customers who couldn't wait, thought Harry, but I thought he just had more stories to tell us. His bowstring would actually be two strands—a single loop of airplane cable, pinched into smaller loops at both ends by sleeve fasteners. These smaller loops would fit into notches at the end of the bow, and the double strand of cable would support a basket in the middle of the bowstring. Rick was fastening the weave permanently by threading the cable ends through steel sleeves and squashing the sleeves in a massive old press, a Skookum-Cooke.

This press had an anthropomorphic appearance, like the futuristic robots of the fifties: an enormous head of solid steel, held aloft by four shiny posts and capped atop its cranium by four canteloupe-sized nuts, a crown. Its body was the hydraulic lift, which opened the jaws of the thing, the head being simply a thick weight that descended to squash the material in the creature's mouth. Rick had rebuilt the Skookum-Cooke himself when it had finally worn out after thirty years in the shop. As a finishing touch, he had painted the press red and green, then sanded off the paint on the brass label so that the raised letters of the Skookum-Cooke trademark gleamed through. It was an awesome machine, said Isaac, and Rick employed it, despite its massive appearance, with some finesse. He talked to us about it as he pressed our cable.

"Old Cooke himself—the guy who invented this thing—came in here one day," he said. Rick pushed a red button on the side of the press, and the jaws chomped down on the sleeve he had placed in a groove inside its mouth. Isaac took a step back.

"The old guy was real happy when he saw this press still working," Rick continued. He'd built it and several others like it up in Portland, and he had never expected to see one in action again. Old Cooke told Rick that after he'd invented the press, he'd gone on to invent Shake 'n' Bake, and was now a multimillionaire. Rick pushed the other button, the green one, and the jaws opened with a hissing sound, the shiny steel pistons exposing themselves. Isaac stood on his tiptoes and peered into the mechanical maw, in which was our cable assembly, nicely squashed. I could see how Shake 'n' Bake might have occurred to Old Cooke.

Rick lifted out the cable and gave the loose ends a stout pull. The squashed sleeve held fast. "Even though he'd gotten rich," he said, "Old Cooke still shopped for his own cable, when he needed some." Less gritty multimillionaires, I thought, might have simply sent the butler down.

When he finished squashing our sleeves, Rick said he needed another day to finish the bowstring. We didn't rush him, though I wanted to. The bowstring was critical. The potency of the springs tended to force things too far one way or another. The basket had to hold the stones loosely enough to let go of them when we fired the catapult, and it had to hold them tightly enough to keep them in the ready position—tricky business for a mesh basket without firm supports of any kind. And Rick also had to devise a way to clip the basket into our trigger. A single eye would do, he said. We were pleasant and seemed leisurely about it, though inside I had that maddening sense that comes from trying to hurry someone by pretending

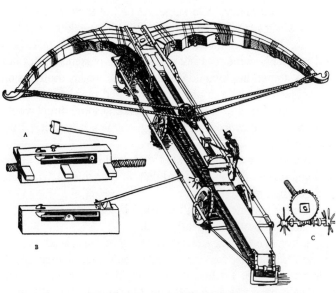

Stonebow by Leonardo Da Vinci

From *The Crossbow*, by Sir Ralph Payne-Galway,
The Holland Press, Ltd., London, 1903

you have all the time in the world. When we finally got out of
the shop, I practically ran to the car.

After that, we dropped Isaac off at home and took the freeway out toward Hayward. When he got the truck to cruising
speed, Harry asked me, as if casually, how many people might
come to this public presentation of the catapult. I didn't know,
I said. Maybe fifty, with the staff of the art center and friends
of friends and everything. I figured we would generate some
interest.

"I think I should be there," Harry said.

"In the audience?" I said.

"No," said Harry. "On stage. As half of the team."

This was a surprise. Harry loathed public speaking, and had from the beginning disavowed any interest in this talk. So I had gradually come to think of the public presentation of our work as something I would do, as a forum for my thoughts on the subject of the catapult. I had planned on showing some slides of our catapult in action, and reading something I would write about the process of building it. "You'll hate it," I reminded him.

"I want to be there," he said. "It wouldn't be right for us to do all this work together, and then for you to stand up there and get the credit."

"I'll give you credit, don't worry," I said.

"I want to be there," Harry said.

I didn't like the idea of Harry being at the talk at all. If I did the talk alone, for one thing, I could plan something. If Harry was there, I'd lose control. He'd say something, and then we'd have to start ad-libbing. Our reasons for building a catapult seemed like a particularly thorny thing to have to extemporize upon. "If you just sit there," I said, "you'll look like some sort of exhibit. If you're there, you're going to have to talk to the audience."

"I want to be there," said Harry. He was worried about how I would portray him. He wanted to control the way he came across. "I don't want to be some character, some funny little guy—like Igor," he said. "Maybe I would be some funny little guy, but at least I'd be up there, being my own funny little guy." And who knows, maybe he was thinking that there would be some glory in it—accolades, applause, whatever. Anyway, he had decided to be there. It was a big mistake.

But I had no real grounds to keep him away from the talk if he wanted to be there. I was the writer, and the project was my idea, but since day one we'd been partners, and it seemed only fair to share the stage with him. But fairness didn't lessen my anxiety any, and I wasn't going to let him nurse any comforting illusions. "There might be a hundred people there, Harry," I said. Harry didn't say anything, seeming to concentrate on the freeway. I wondered if I could write a script for the whole event, and if I could, if Harry could stick to it. "It should be fun," I warned him. He didn't answer. "You really want to do this?" I said. "Yup," said Harry. I didn't argue with him about the lecture, but from that moment I had a bad feeling about it.

We drove into more suburban territory, and right off the freeway we found one of those big commercial hardware stores, a sort of hardware mart, the kind with bags of mulch stacked in a fenced-in area of the parking lot. We planned to do some odd shopping there for something they obviously didn't stock. We were looking for catapult trigger parts, and we needed a store with a lot of different kinds of stuff, some of which might do.

Harry had thought about the trigger since the beginning of the project. We needed to bring the power of the springs to a point at which we could engage it fully and immediately, the way a shutter engages light. To do this, Harry had come up with an archer's trigger, a sort of steel index finger, crooked to catch the bowstring, and hinged at its base to release it. The thing would sit on the back of the beam like a switch, its hinge welded to a plate bolted to the beam. We planned to block the device behind its hinge to prevent premature firing, and Harry thought we could attach a rope to the block, and release the trigger by tugging on the rope, pulling the block out of the hinge, and allowing the trigger finger to pop off the bowstring.

The trigger posed only one real difficulty. It would have to hold so much force that it seemed likely to hang up and refuse to fire, the pull of the bowstring jamming the trigger mechanism. Just as likely, though, was the opposite effect of the force—that it would overwhelm the mechanism and make the trigger slip. The trigger had to provide exact control at the moment of decisive change. Harry had been studying his woodcuts and worrying about this critical part.

So we walked through the aisles of the big store, pushing a shopping cart and looking for something we could use as the main pin, or finger, for the trigger. We had to disregard what anything in the store was actually called, and search for the proper set of characteristics. The thing had to be steel, and it had to be crooked at pretty much a right angle, said Harry. We could probably have bent a steel rod at David's, but the thing's simplicity seemed general enough that someone would have produced something like it that we could use—a ready-made. Besides, as far as Harry was concerned, we were already beyond our limit when it came to leaning on his brother-in-law, who, despite the fact that we had paid him, was going to expect return favors from Harry, faux-graniting stereo speakers or something like that.

Harry found a fireplace poker that more or less embodied our set of characteristics, and we stood in the aisle, arguing the finer points of the design of this poker. It had a crook, though not a right-angled one. Harry argued that a completely perpendicular crook would grab the bowstring too fully and cause the mechanism to jam. This poker's crook was too far off, I said; it didn't look like it would hold at all. Besides, I thought, the thing looked stupid: black fake wrought iron with useless Gothic horns and brass plating on the handle. I didn't want anything so stupid-looking on our catapult. It would work, Harry said, hanging on to the poker, so I abandoned him, and pushed my shopping cart into the construction hardware department, hoping to rescue the appearance of our catapult before Harry got to the cashier with the poker.

I walked until I found an obscure bin full of obscure right-angled steel rods, J-bolts, the sign said, parts just as we imagined, except that these were threaded for a nut on the end of the long part of the J. The J-bolts were foundation tie-down pins, things that held your Walnut Creek estate to its concrete pad during an earthquake, I thought. I took one, and found

Harry waiting in line with his poker behind a guy who looked like a divorce lawyer on his day off, a roll of garden hose under his arm. I showed Harry my J-bolt, and we debated the merits of our respective finds.

"That thing's going to hang up when we fire the thing," Harry said.

"Well, yours is going to look stupid," I said. The guy with the garden hose was glancing surreptitiously over his shoulder at us. Harry said we'd probably have to modify the crook of the J-bolt to use it, and I asked how we were going to weld anything to the brass finish on the poker. Finally Harry agreed with me. He went to put the poker back, and I exulted silently in the cash register line. For one thing, the tie-down pin cost about twenty bucks less than the poker.

Suddenly, the guy with the garden hose turned around and said, "Do you mind if I ask what you're building?" I was in a good mood, so I told him. We were making a catapult, to throw stones, and this pin was part of the trigger.

"That's what I thought," said the gray-templed guy.

"Actually, we're going to lay siege to the Hilltop Mall," I said. That didn't faze him either, though, as if people shot stones at the mall every couple of weeks. We both smiled perfunctorily and he turned back around, leaving me with the feeling that I had gone native in our alien pursuit. I felt deprived of my irony, there in the cash register line, and doubtful, suddenly, of the transcendental aspect of the project, that we were just observers in the world of gun-freaks. When my co-conspirator returned, I pointed at the garden hose guy's back and rolled my eyes. The guy just didn't get it—but maybe nobody ever would. "Back to the weapons workshop," I said, as we walked to the truck with our new trigger finger.

After that we were into Thanksgiving week. We had planned to shoot the catapult on the first Sunday in December. We had to give our talk, I reminded Harry, only a week after that. I was hoping to shake his resolve, but Harry remained obstinate. We were watching for rain, which, after months of dry weather in northern California, usually arrives in late November, the sky boiling with low clouds for months afterward. I could imagine us, cold and wet, in plastic ponchos, mud up to our ankles, trying to set up the catapult on the Headlands in the rain, like a scene from the Great War. But the clear, dry weather held. The storm track stayed north over Canada, where it had been all summer. The weathermen were already talking about a third year of drought.

On the day before Thanksgiving, I went downtown to David's to pick up the gussets. Workers in their cars jammed the streets, their headlights already on against the early dark. The welding shop was empty and bleak, the welders off early for the holiday weekend. David was waiting for me in the little office, and on his desk lay the parts he'd made for us—the gussets, cut to match Harry's cardboard templates, some pieces for our trigger, and our springs, which he'd modified. The stuff looked thick and dense there, like pieces of chocolate. David greeted me laconically, but as I wrote him out a check, he brightened a little and told me a story, the tale of the demise of the welding shop. The old guy, Bachman, had created and run the place almost single-handedly for nearly a half century. He had worked until he was past seventy, and then had made an attempt to pass the business on to his son. But the son had never wanted to work for his father, though he'd done so in any case for years. He had never taken to the sooty work of the place, and by the time he was finally allowed to take over, he himself was thinking about retirement. So the soaring real estate values in the neighborhood proved providential, and when the Japanese group offered him the two million for the shop, he took it. "Two million dollars," said David, raising his hands to the steel shelves above the desk, the ring binders and parts catalogues, "for this." So Old Bachman wasn't speaking to his son, but who could blame the guy for selling out? Of course, it meant that David would have to find another job. He had applied at The Good Guys, he said. I wrote my check to this vanishing concern, paid David, and wished him luck selling stereos. He'd probably be good at it, I said—he *was* a good guy, anyway. David smiled wanly.

I had to make a couple of trips to the car, loading the steel. The shop's neon robot welder shed orange light over the sidewalk. It was getting late, I thought, as I lugged the stuff out. At least these were our last few parts. It seemed odd that they might be among the last steel pieces to come out of Bachman's—fifty years of modern welding, concluding with this odd flourish of catapult parts.

I drove to Harry's, across the bridge through heavy traffic and east into what was already night. We were getting down to it, I thought, down to essentials. The elaborate holes in the springs hadn't been necessary—David had cut open the end holes for us with his suicide saw, making simple notches for the loop of the bowstring.

Harry wasn't home yet, so I drove to his shop, and found him still at work, trying to finish stuff up before quitting for the holiday. As usual, he was covered with a fine mist of sprayed paint, this time pink. I showed him the steel I'd picked up at David's. We looked at it like the two weary workers we were. One of the pieces, a flat steel rectangle drilled for bolts, would be the base for the trigger, and it had to be bent. It had to sit like a saddle on our beam. Harry said he thought there was a metal shop nearby where we could get somebody to bend it for us, and when he closed up, we drove past the place, and found its big doors still open, its lights still on. We stopped and took our steel plate to the door.

Inside the workers were standing around holding cans of beer and murmuring in Spanish. They had obviously quit for the day, and they eyed us with tired annoyance. Not put off in the least, Harry explained what we needed, pointing out the lines on the steel where it had to be bent. When he finished talking, nobody moved. These guys were blackened from their work. It was already Miller time. They just stood there, holding their beers, and looking at this thick-shouldered, pink-painted man with his piece of steel.

"Come on," Harry cajoled them. "We'll pay you money, we'll pay you money."

Finally one guy shrugged, put down his beer, and took the steel from Harry. He walked over to a long set of steel jaws, a tool called a metal break, and laid the piece in it. He pulled a lever and the jaws folded our steel like modeling clay. Then he crimped it again on the other side, and handed it back to me. The plate was now a rigid, heavy, three-sided box in my hands, warm at the corners from its bending.

"How much do we owe you?" I said.

"Nothing," said the guy with a level gaze. "Not a thing."

CHAPTER 19

Tube Alloy

In the seventh year of the twentieth century, the New Zealander Ernest Rutherford and his protegé Ernest Marsden thought of shooting the alpha rays produced by radioactive elements through gold foil. To their confoundment, they noticed that instead of passing through the foil, some of the rays bounced rather radically off it. The Kiwi genius was flabbergasted. "It was almost as incredible," he said, "as if you fired a fifteen-inch shell at a piece of tissue paper and it came back and hit you." His rays had hit some enormously strong thing. Rutherford deduced that the atom had to be mostly empty, its power concentrated in a nucleus, "the seat," he thought, "of very intense electrical forces."

Rutherford's artillery simile and his use of the word *electrical* to describe what he'd found sprang from tenets of the nineteenth-century scientific mind, and just glimmered of things to come. His was the most brilliant of laboratories of the build-your-own era, and he had it essentially right. But he was like the tourists at El Capitan, measuring the monolith by stories, and, no wonder, coming up short. For here in the nucleus of the atom was absolute power, power as absolute, anyway, as the best mind of the last century could gather. But science would proceed. Rutherford's rays, it turned out, had not bounced off the nucleus, they'd wrapped around it and surged away, like a comet, like a rock in a sling.

Ten years later, Rutherford was again engaged in the atomic equivalent of finding out how a house is made by blowing it to pieces and looking at the bits. He was bombarding nitrogen with alpha rays when he noticed that he was knocking loose big particles, bits he could by then compare to hydrogen nuclei, singular and negatively charged. He called them protons, and proposed them as a basic constituent of matter.

By 1920, he was proposing another, the neutron, which his protégé James Chadwick located twelve years later. Rutherford recognized at once that as an atomic projectile—as a sort of cannonball—the neutron might be ideal. Its neutral electrical charge would carry it unaffected through the power fields of atoms. Plus neutrons were massive, at least compared to the current ammo, alpha rays. They could do far more damage than alpha rays did, to say the least. Later the physicist I. I. Rabi described this atomic collision in planetary terms. "When a neutron strikes a nucleus," he wrote, "the effects are about as catastrophic as if the moon struck the earth."

In 1938, as Hitler planned the war that would execute his Final Solution, two chemists at a state laboratory near Berlin shot neutrons at uranium, and found something quite weird and momentous in the rubble. They found barium. It was as if you blew up a TV and found a vacuum cleaner in the wreckage. Cautiously, they worded and published their findings. The chemists, Hahn and Strassman, were wary of drawing any conclusions in the matter, which had a revolutionary implication about the physics of atomic structures. But soon the international community of physicists understood the implication—that the Germans had broken apart the unwieldy nucleus of a uranium atom, splitting it into lighter atoms like barium, and had done so in collisions so massive and complete that the flying debris might similarly collide, setting off a kind of atomic fire. This was fission, named, as Richard Rhodes notes in his exhaustive chronicle of the atomic bomb, by a perversely applied analogy to the subdividing of cells in the process of life. The splitting of the atom also set off a spectacular and rapid chain of thoughts that fleshed out the implication of fission in the context of Hitler's campaign. Within a week, a rough sketch of an atom bomb, a weapon for use against the Nazis, appeared on J. Robert Oppenheimer's blackboard at Berkeley. And within one year, a proposal to build such a weapon sat on the desk of the president of the United States. Roosevelt was pondering his response to this report when the Japanese bombarded Pearl Harbor, setting off the war.

As the war and the development of the atom bomb proceeded, neither the thing nor its materials could be referred to directly in public. So its names proliferated, the true one—like Yahweh's—top secret. In his notes, the Supreme Court justice Felix Frankfurter referred to the bomb simply as X. The British concocted the nonsensical spy-term tube alloy. Secretary of War Henry Stimson called the bomb S-1. Robert Oppenheimer and the scientists at Los Alamos named it the gadget, an intimate term suggesting its size and intricacy. They referred to the weapon by this cute name even in sessions planning the actual atomic bombing: "Place first gadget in center of selected city" was one strategic directive. Paul Tibbets, the pilot who dropped the bomb called Little Boy, referred to the weapon among his crew with a hot dog's derision. He called it the gimmick.

Tibbets, incidentally, was thirty years old at the time of his mission, and was called the Old Man by his crew. Tibbets had ordered his mother's name, Enola Gay, painted on a B-29's fuselage on the day before the mission, in advance of what would come to be called the photo opportunity. But it wasn't actually his plane. As commander, he had commandeered the aircraft, and its pilot-cum-copilot Robert Lewis was outraged to see *his* airplane bearing this other man's mother's name. At Los Alamos, X stood for explosives. Conventional high explosives triggered the atomic bomb, uniting in their blast the parts of the nuclear critical mass in the heart of the weapon. The scientists at Los Alamos divided the tasks involved in engineering the bomb, and Oppenheimer called the explosives group Division X. The head of the X division, chemist George Kistiakowsky, was quite comfortable around things that blew up. Using a dental drill, he bored into fifty-pound blocks of high explosive to fill inner cavities that might have disturbed the regularity of their detonation. "You don't worry about it," he said. "I mean, if fifty pounds of explosives goes off in your lap, you won't know it."

Kistiakowsky's view of high explosives was that they could be "precision instruments" rather than simply "blind destructive agents." When the scientists decided to clear a slope of trees to make a ski run, Kistiakowsky wrapped the trunks of the trees with high explosive, set it off and blew them cleanly down.

He and Oppenheimer rode horses for recreation at Los Alamos, a place that seemed to the Easterners like a dude ranch and where the unofficial uniform for civilian workers was cowboy gear. (Years later, the chemist Hans Bethe would appear in a documentary on the project, still wearing his string tie.) Kistiakowsky bought a fierce black horse from Oppenheimer. Its name was Crisis, testimony to Oppenheimer's love for critical situations. "All things which evoke discipline," he had written, ". . . ought to be greeted by us with profound respect." For Oppenheimer, crisis provided definition not possible under ordinary circumstances. The crisis of the war had borne their mission, for instance, and even the application of nuclear fission for a bomb might not have occurred in a context other than the war's. In any case, Kistiakowsky broke in the horse called Crisis, riding for miles over the mesa on Sundays, his one day off.

The arming of the atomic bomb was its central technical challenge. Initially, engineers proposed that the bomb be spring-loaded. Springs would bring together two halves of a uranium sphere, initiating a chain reaction. Later the concept of a gun-bomb replaced the idea of a spring-loaded bomb, recapitulating the historical step from the catapult to the cannon. Seemingly designed by erotic analogy, the gun-bomb exploded upon mating: a cannon would shoot a plug of U235 into a radioactive sphere, completing it and igniting the process called fission. Little Boy was such a gun-bomb, its sphere welded to the muzzle of an unrifled gun, the plug shot into it over Hiroshima. But even before its single use, the gun-bomb was obsolete.

Little Boy used an enriched uranium isotope, U235, for its critical mass. U235 occurs naturally in less than one percent of the volume of natural uranium, and is "hotter"—more radioactive. A critical mass of natural uranium would have weighed tons, too much for an airplane. So U235 made a critical mass of a manageable size. But the new reactor by-product called plutonium turned out to be even hotter than U235. A critical mass of plutonium was only about the size of a small orange, but proved so hot that no gun could fire a bullet into it fast enough. Even with a muzzle speed of three thousand feet per second—the fastest available—a gun couldn't insert a plug into a plutonium core quickly enough to achieve a clean burn. The premature reaction would cause a meltdown in the core before the plug—even traveling at such a speed—could be completely inserted. A colleague in the X division had proposed implosion, instead of insertion (rather than shooting a single plug into the core, the detonation would set off simultaneous explosions around it, squeezing the hollow center into a unified critical mass), and Kistiakowsky took up the challenge.

He had to do the work under pressure. The air force had been systematically bombing Japanese cities, "with the prime purpose in mind," read their directive, "of not leaving one stone upon another," and in the spring of 1945, informed the Los Alamos team that by January 1, 1946, there would be no suitable targets—intact cities where the damage might be measured—left in Japan.

To make implosion work, Kistiakowsky had to perfect exploding lenses, shaped charges that could direct a blast with exacting precision. Molten components of these explosives had to be poured into a mold, and then cooled precisely to attain the desired consistency. The scientists, wrote Kistiakowsky, "sat over the damned thing, watching it as if it was an egg being hatched." The blocks of high explosive then had to be machined—band-sawed—to almost impossible tolerances, measured in thousandths of an inch. Kistiakowsky nearly blew himself up on several occasions, once igniting a small forest fire.

Implosion required other crucial developments. The X researchers had to use a new device, an IBM card sorter, to calculate the vast numbers involved in the mechanics of the bomb's internal shock waves, and had to perfect split-second X-ray photography techniques to observe the explosions as they occurred inside the bomb. Such difficulties nearly overwhelmed Robert Oppenheimer, who had to be talked out of resigning when the technical feats involved in implosion proved critical to the project. Kistiakowsky remained confident, though—even when things looked the worst, he bet Oppenheimer $10,000 to ten dollars that his lenses would work.

They did. The first nuclear explosion occurred within a second-generation weapon, the implosion bomb known as Fat Man. The day they hoisted Fat Man to its tower at the Trinity Site, Little Boy's untested gun-bomb components were already on their way to Japan, shipped first by air to the Hunter's Point Naval Shipyard in San Francisco. It was being loaded on board the cruiser *Indianapolis* as the Fat Man went off at the Trinity Site. The *Indianapolis* then sailed out of the Golden Gate, under the gaze of the lookouts manning the gun batteries on the heights called the Headlands.

Meanwhile, at Los Alamos, Fat Man had produced an explosion beyond the expectations of most of the atomic scientists—the blast was equal to 18,600 tons of TNT, four times the official prediction. His expertise with explosives notwithstanding, Kistiakowsky stood unprotected outside a bunker ten thousand yards from ground zero—more than five and a half miles away—and was knocked off his feet. He had predicted an explosion about one-twentieth of the size of the Fat Man blast. When he got up, he claimed his ten dollars from Oppenheimer, who, it turned out, didn't have the cash on him.

Nobody seemed to have anything to say that was adequate to the occasion of the first atomic explosion. Many of these scientific witnesses simply offered vague superlatives describing their own muteness. They were "flabbergasted," they said, by the "overwhelming" flash, a thing of "unbelievable brightness," and "completely breathtaking." Others, still vague, at least convey a sense of honest fright. "You would wish it to stop," wrote the physicist I. I. Rabi. "It looked menacing. It seemed to come toward one."

Seemingly more descriptive were the empirical, rather geometrical reports of the scientists. "The stem appeared twisted like a left-handed screw," said one. Another noted the "several spikes that shot radially ahead of the ball below the equator" and the "wide skirt of lumpy matter" ahead of the shock wave. Others almost missed the event altogether. Enrico Fermi busied himself during the blast by scattering bits of paper in the shock wave, by which he hoped to measure the yield of the explosion. "I could observe very distinctly," he wrote, "and actually measure the displacement of the pieces of paper." Laura Fermi noted that her husband was "so profoundly and totally absorbed in his bits of paper that he was not aware of the tremendous noise."

Frank Oppenheimer lay on the ground next to his brother Robert as they witnessed the blast. Frank recalled the sound of the explosion, the thunder that "never seemed to stop," echoing and reechoing off the desert stone. But he couldn't exactly recall what he and his brother had said at that moment. "I think we just said, *it worked*," he says in Jon Else's film, *The Day After Trinity*. "I think that's what we said, both of us. *It worked*." Robert Oppenheimer, for his part, attempted literary allusion, writing later that he thought of Prometheus, of Vishnu as Death, the many-armed destroyer, and of Alfred Nobel and "his vain hope, that dynamite would put an end to all wars."

A woman and her sister were on a highway some forty miles from the explosion, and the sister, who had been blind for years, remarked at the flash. The Italian physicist Emile Segré, five miles from the blast in the cold desert night, described its heat. "It was like opening a hot oven," he said.

The secretary of war's emissary at the site, a man named George Harrison, conveyed the news of a successful explosion to Washington in rather different terms: "Operated on this morning," read his telegram. "Diagnosis not yet complete but results seem satisfactory." Later he added, "Patient progressing rapidly," and finally detailed his medical terms into a full-fledged obstetrical analogy: "Doctor has just returned most enthusiastic and confident that the little boy is as husky as his big brother," he wrote. "I could have heard his screams from here to my farm."

General Dwight Eisenhower heard this cable read aloud after a dinner party with the secretary and was disgusted. "The cable was in code, you know the way they do it," he wrote. "*The lamb is born* or some damn thing like that." Eisenhower opposed using the atomic bomb against the Japanese, viewing their surrender as imminent in any case, and dubious of the judgment of history: "I hated to see our country be the first to use such a thing," he wrote.

President Truman left no record of his formal authorization

to drop the bomb on the port city of Hiroshima, which means Broad Island Castle in Japanese. In his diary, Truman retreated into the passive voice to describe the decision: "The weapon is to be used against Japan between now and August 10th." The passive voice has the luxurious advantage of not requiring a subject for the verb, making it seem almost as if the bomb had a mind of its own. Copilot Robert Lewis had a similar thought, hearing over the airplane's intercom that the bomb-bay crew had finished arming the weapon. "I had a feeling the bomb had a life of its own now, that had nothing to do with us," he noted in his journal.

The words of the witnesses in Hiroshima are particular. A woman trips over a man's decapitated head, and involuntarily cries, "Excuse me." A man carries his eyeball in his hand. An old woman's skin hangs from her body "like a kimono." Three charred schoolgirls sleepwalk away from the devastation, holding their burned arms in front of them "like kangaroos." Many people have no faces, no hair—the witnesses can't tell the fronts of these naked, burned people from their backs. Crowds of the burned drown in the city's water cisterns.

After the war, the army and navy awarded the team at Los Alamos with a symbolic flag. The banner bore the words *Army* and *Navy*, along with some stars and a laurel wreath—signifying aspiration and attainment, apparently, and within the wreath, the letter E. E stood for excellence: this was the army-navy E. The scientists themselves convened as the Association of Los Alamos Scientists, the name suggesting an archaic cry of woe, only incidentally, perhaps, by its acronym—ALAS. A somber Robert Oppenheimer addressed the group at the war's end, as he took up his unsuccessful work of changing the world to fit the bomb, of convincing the U.S. government of the enormous scale of the atomic bomb, so far beyond previous frames of reference that it required remaking our sense of nature and nations. "In the actual world and with the actual people in it," he told ALAS, "it has taken time, and it may take longer, to understand what this is all about."

By then Oppenheimer was being disappointed. For him, as for many of the men at Los Alamos, working at Los Alamos had been the best time of their lives. There had been a community. They had had around them the most brilliant of their colleagues, assembled from all over the world, and, given their fervent purpose, they had banished for the most part the jealousy and personal competition that such a talented assembly would engender under ordinary circumstances. Their differences had dissolved in their task. They had tapped the secrets of the atom, which like some mother lode had lain there in the deep, and they had built the most powerful device ever composed. Their minds had offered them the absolute all their lives, and they had seemed at Los Alamos to be taking the absolute in hand.

But it might have occurred to some of the scientists at Los Alamos that it is ultimately the target that provides a bomb with its power. The army's directive to drop the bomb had to have two parts, one technical and one geographic: place gadget in center of selected city. An explosion itself is merely a potent void, a forceful exclusion, a greedy nothing, measured in terms of the hole left in the world where the bomb went off. Looking at the explosion abstractly, in the desert at Los Alamos, they mistook the ground for the figure, the cipher for the sum. Without the city, the bomb was a cosmic spark, interesting and useful enough in its place in the stars, but on earth in a human city, evil. This stunningly obvious conclusion somehow escaped many of the Los Alamos scientists, some of whom failed to reach it even after they heard the reports of the Hiroshima witnesses. By then, they were immersed in what Richard Rhodes has called the "technological imperative, the urge to improvement even if the objects to be improved are weapons of mass destruction."

This urge I understood. My grief over Hiroshima I felt in

part as guilt for my admiration of Los Alamos, for that sense of community, mission, exploration, control and danger. I mentioned my feelings of implication to Harry. He said I need not feel any special guilt. Anyone would feel that way, he said. That was *the* tragedy.

CHAPTER 20

Uncle Frank

Robert Oppenheimer was an elder brother, and couldn't duck when the time came. He strode off the mesa in his jeans, walked onto Capitol Hill in a suit, and got his block knocked off. He may have figured he had saved the country once *with* the bomb, and that he was going to save it again *from* it. But the country would not cooperate in this second rescue. Oppenheimer opposed the up-and-coming Edward Teller and his big new bomb, telling him to keep his shirt on. And he advocated the impossible, the free exchange of nuclear secrets with the Russians. He had wanted Truman to tell Stalin about the Trinity Site blast, before Hiroshima. He trusted in such openness; as a scientist, he'd seen science evolving via shared information and debate. More critically, he knew that there was no real alternative to sharing information. These secrets would not keep. Fission and the other processes operating within the atomic bomb were real, real the way a mountain is real, however remote. "They are found," he said, "because it is possible to find them."

So withholding nuclear knowledge from the Soviet Union offered a temporary advantage, at best, he felt, one not obviously outweighing the diplomatic benefits we might reap by our generosity. But the nation, braced for an enemy, refused him, made him a witch for the witch-hunt. Under pressure, the Atomic Energy Commission revoked his security clearance, in effect banishing him from the arena of policy. So Fate, it might be said, captured Robert Oppenheimer. He had made himself its special instrument. Denied that use, he lived bitterly and not long.

Robert's younger brother Frank was if anything more vulnerable to the tarring that he also received in Washington. He had less fame and less authority to shield him, and the central charge against him, of being a member of the Communist party, was true—he had joined the party in the late thirties in support of the Spanish Loyalists against Franco, and had resigned when he went to work against the Nazis in the government lab at Los Alamos.

"During the war," Frank said later, "it seemed enormously important to me that America develop an atomic bomb as quickly as possible and before anyone else did." At Los Alamos, Frank monitored wind shifts and car traffic around the base during the Trinity Test, among other things. He made his own experiments on the site as well. He set up wooden boxes and pine boards at various distances from ground zero, and filled the boxes with excelsior—shredded pulp—to simulate Japanese housing. His results implied a mile-wide circle of charred boxes around such an explosion. It was an experiment both disinterested and alert to the larger purpose of the event, the sort of experiment Frank would make his entire life.

But not as a theoretical physicist. Summoned before the House Un-American Activities Committee in 1949, he spoke freely of his own past, but refused comment on a list of names proffered by the prosecution. Technically he was in contempt, but the committee was after bigger fish, and Frank was not charged. He was, however, deprived of academic work, forced to resign his professorship at the University of Minnesota and denied jobs elsewhere. But in the face of this, Frank showed the resilience that may be the special gift of younger brothers. Frank retreated in grief with his family to Colorado, to Pagosa Springs, where he took up cattle ranching eighty miles from the nearest hardware store, in a place the neighbors called the Basin. On the ranch, Frank said later, he learned "about getting things stuck and getting them unstuck."

It would be enough for anyone to expect that Frank Oppenheimer would come, in the end, to like cattle ranching. But he seems to have loved it from the beginning. For one thing, it was his own in a way that, with such a brother, nuclear physics could not have been. And it turned out he didn't have to leave his old experimentalist self behind when he came to the ranch. Far from it, he had to improvise more than ever, making his own tools, plowing his own fields, and when a heifer delivered or the cows got sick from eating larkspur, doctoring his own animals.

But he wasn't doing science, as the FBI found when they checked with his neighbors. The world's curse—and its blessing—is that it won't go away. The feds came to visit the Oppenheimers, bearing the same list of names, and asking the neighbors if there had been any odd experiments going on at the ranch. "You know," Frank would say later, "the mad scientist doing terrible things." But the time for doing terrible things was over. "Now the making of atomic bombs seems repugnant and evil to me," he would say in 1960. And the neighbors liked Frank. They elected him chairman of the local phone company, and when the launch of the Sputnik convinced them that they needed more than one science teacher at the high school, they hired him to teach.

Odd, what he must have felt then. He had said good-bye to science, and here it was again, transformed. For Frank, new by then as well, recognized science as a larger inquiry than it had seemed at first. He had realized that the question of *what one knows* stood upon the question of *how one knows*. "If you are going to know about nature," he said, "you have to know about how people react to and feel about nature." In the classroom this distinction was not the abstract matter it might have been elsewhere—it arose urgently enough out of the challenge of wanting to be a good teacher. "I wanted you to understand the things I enjoyed understanding," he told his graduating class, "such as why a star got hot and stayed hot." He needed to teach fifth-graders to think about physical optics and Spirogyra, and, as he confessed in a PTA address, he didn't know how. "I can give them a problem to work out and say *Think*," he said, "but this procedure is about as effective as saying *Wiggle your ears*."

This mystery of how one could convey "the ability to fashion a new idea" brought Frank off the ranch and back into the world, where he fell back upon what had always been the chief attraction of scientific inquiry for him, the experiment. His life as a rancher had only increased his enthusiasm for trying things out and seeing how they did. So as a science teacher, his method was vigorously hands-on. And as this rural school board was short on funds for laboratory equipment, he took the students to the town dump, and let them experiment by making things out of junk. He taught them the principles of mechanics and thermodynamics with old car parts. Two of his students won Colorado Science Fair firsts with experiments made from junk.

Soon he was at the university again, amassing a "Library of Experiments" for his students there. Amid this collection the students could play with the concepts taught in lecture. Somewhere along this line, Frank recognized that such a collection of exhibits might address the general population, not just students, and might be an institution on its own. Unique but analogous enough to a museum to be called one, Frank's institution would bear as great a similarity to a laboratory or an astronomical observatory. For the exhibits in this museum would be means, not ends. The visitors would observe the world itself by means of the exhibits, observing at the same time just how it was one went about observing the world in the first place.

Frank got a Guggenheim Fellowship to study museums in Europe, and then moved to California, Land of Experiment, to start his own. He got a grant from a San Francisco foundation, and a lease from the city of a dollar a year for an elaborate neoclassical barn, called by its booster builders— who put the place up for the Pan Pacific Exposition of 1915— the Palace of Fine Arts. Outside the place was pink, enormous, rotund. It was superficially worldly, profoundly provincial. Maidens bearing amphoras upheld its porticos. Inside it was all business and ironwork, cavernous, girdered and elaborately strutted. It had been an airplane hangar, among other things, and would always resemble one. Frank and his wife, Jackie, called it the Exploratorium. Somebody later likened the big space to Frank's mind.

One of the exhibits at the Exploratorium seems to have come directly from Frank's auto junkyard days: it is the drive shaft and axle mechanism from a car, with the casing over their intersection composed of clear Plexiglas, so that one can see how the power is transferred to the wheels from the spinning drive shaft by means of the differential. Transferring this power would be simpler if the car didn't have to go around corners. When that happens the wheels must move at different speeds, the one on the outside of the curve turning faster to complete that longer distance while the inside wheel completes its shorter arc. So the gears in the differential must somehow be made to ration power on that slowed inside wheel. When I told Harry about my surprise in finding this out, he looked at me like I was crazy and exclaimed, "Of course! Otherwise the car would corner like a chariot!"

The first time I went to the Exploratorium, I looked at this differential exhibit for a long time, mainly because there were kids playing with the thing. One boy would grab one wheel and one the other, then, because the differential would choose which kid was putting up more resistance, it would effectively elect the winner. One of the boys would be thrown off. Somehow, watching the kid get thrown off of the thing made the differential mechanism easier to understand.

As with the Plexiglas casing over this differential, so with the rest of the Exploratorium. There was nothing behind the scenes. Frank said he didn't want anybody leaving with the feeling, "Isn't somebody else clever?" In a documentary on the place, the staff argues about money in front of the cameras in a way no company would dream of. Frank even put the machine shop behind Plexiglas walls so that the visitors could observe the designers assembling new exhibits, there in the watchworks of the museum itself. The open technology was the whole point, Frank said, "to make it possible for people to believe they can understand the world around them." A woman wrote to him that after her visit to the Exploratorium, she went home and wired a plug on a lamp cord. Telling the story in the film, Frank notes proudly that there was nothing at the Exploratorium about lamp cords.

Frank Oppenheimer and his brother Robert shared the same goals after the war. They had witnessed the bomb at Los Alamos, and they needed to make the world aware of the enormity of that event. Even this tiny first blast was so far beyond the pale of previous conceptions that the brothers feared it would not be recognized in its scale. People would think of it as just another weapon, just another rock, and begin stockpiling arsenals. But Robert Oppenheimer's frontal assault on government policy was like Prince Edward's sentimental charge against the Londoners, the stoners of his mother—weighty with self-esteem and the need to redress, and doomed.

Frank Oppenheimer took a longer approach to changing the world to fit the bomb. "The basis for social change, as well as technical change," he said, "is in understanding how nature behaves and how people behave." If he and others could create confidence in that kind of understanding, he reasoned, "there's some chance that we won't blow each other up, and that we can have a decent society." In this quest, he was hopeful.

When later in life someone would protest to him, "But Frank, we live in the real world," Frank would answer, "No we don't. We live in the world we made up."

And not only do we live in a world we make up, Frank believed, but we make it up so in a certain describable way. From the world of sensory data, we recognize patterns, and further, we recognize patterns among those patterns, "called theories in physics," said Frank, "or compositions in painting." Both Newton's Theory of Gravity and Picasso's "Guernica" constitute a synthesis of experiential patterns, and both involve a process of selection. And both are valid, Frank thought, because they lead to the revelations of things that are happening, though not yet perceived. "Many artists' sketches," Frank said, "simply portray or describe a newly discerned pattern."

So if the Exploratorium was born by analogy as a museum, whether it was a museum of art or a museum of science might be debated. Frank paid both artists and scientists to design the exhibits. His enlarged sense of scientific inquiry, his leap from the objects of perception to the question of perception itself, resolved science and art as the twinned manifestation of a single function, part of the mechanism by which the consciousness of the species evolves.

This theme of resolving patterns was the leitmotif of Frank's existence. When, late in his life, he placed the discovery of widespread unity in nature among "the most elegant and satisfying achievements of science," he was speaking of his own lifelong pleasure in discovering such unity, as well. He recalled his excitement over high-altitude balloon experiments, where he found not just hydrogen, but atoms of all the elements in the stream of cosmic rays from outer space.

By then Frank had come to his own complex, contradictory-seeming fate, and he was grateful for it, a nuclear physicist managing the day-to-day details of life at the Exploratorium. Among the various profundities he offered in an interview two years before his death in 1985, he complains against the basic design of the vacuum cleaner, that cumbersome and howling machine. "I'm sure it's not the economics of it," he says. "It's just that nobody paid any attention to the feel of it, or to what it was doing to other people."

CHAPTER 21

Pot Metal

Harry and I went down to the Exploratorium on the day after Thanksgiving to do the final bits of machining on our catapult. We were coming down hard on deadline by then. Thanksgiving Day had intervened, and with Harry's whole family coming over for the day, we didn't think it would look right for us to be working on a weapon on the back porch. Sara and I had dinner at Harry's, too, with his mother, Sandra and David, along with a couple of cousins and their husbands and kids. Susan made a huge meal, and the feast was a raucous affair, as usual, lasting all day, and ending late with the uncles wrestling the boys in the living room.

At one point before we ate, everybody trooped out to the porch to look at this thing that Jim and Harry were making. The catapult lay on the porch like the prostrate skeleton of a huge bird, and the family gathered around it. Harry's mother, whom we call Bubbi, didn't like it. Cousin Bill thought it was cool. David wondered whether it would work. Some of the women glazed over. The whole premise was stupid to them, Harry said later.

I had a friend at the Exploratorium, an exhibit designer named Ned. Ned said we could come down there after hours and use the tools. Harry didn't want to go. He had always protested against any kind of expert advice or help. Once I told him that at a dinner party I had spoken to a physicist about the mechanics of springs, and Harry complained that I was going to wreck the project. Harry didn't want to know about how we might make a state-of-the-art catapult. A state-of-the-art catapult would make us look like goons, he said. He wasn't sure that we didn't look like goons already.

But I was nervous about the last bit of delicate work, the tooling of the trigger. Almost everything we'd done so far had been gross, so to speak, big enough so that our mistakes wouldn't prove fatal to the project. The finest work we'd simply hired out to Rick at Feeney's, but this last stuff was still fine enough, I thought, and we would have to do it ourselves. We would need good light and good tools. It would be precision machining, I told Harry.

"Don't make me laugh," he said. "Everything we're doing is crude—get that through your head." I didn't know what precision machining was. We could make a revolver with precision machining. We're making a machine to throw rocks, he reminded me.

Still, I could tell I was making him nervous just by posing the questions. How would we weld such small parts? I asked. Was it going to be harder than welding big parts? What kind of saw would we need to use on the pieces of the trigger? Harry seemed to dismiss my anxieties about these details, but I could tell my questions were sticking with him. Often after I needled him this way, he would come up with my own idea a day or two later, advocating it as if I had never said anything about it. But at this point, I didn't have time for that, so I told him I'd buy him dinner and that we could see Chico MacMurtie's robots when we got there. Harry hardly ever went out to eat in San Francisco, and I had already told him about Chico's robots, weird skeletal creatures, one of which knew how to throw a rock. So under the dual stimuli, Harry agreed. But he knew about the open shop at the Exploratorium, though, and didn't like it. "Are we going to be on display down there?" he asked. I told him we'd be going at night—the place would be closed.

Harry remembered the Palace of Fine Arts from his earliest childhood, when he lived in that Marina neighborhood. He remembered it as mostly abandoned, fenced and off limits, the big space a garage for Laurel-and-Hardy era city trucks, the decor going to ruin. A big piece of Corinthian capital nearly fell on his father, Harry told me, when the two of them were walking their dog, Mugsie, on the grounds of the place. The big chunk fell about a foot from his father, and stuck into the mushy grass. But the incident that Harry remembered most clearly about the old Palace of Fine Arts was a worse one, one of his earliest memories, of being attacked by a swan near the pond there. The thing gave him a painful bite, and he had never forgotten it. "They swarm on you, those swans," he said, thirty-five years later, as he drove his truck to the Exploratorium.

It was Friday evening, and the parking lot was deserted. Harry and I pushed our way through the big green doors of the Exploratorium's staff entrance, and found ourselves blinking in the light among the machines, our arms full of the steel stuff we had stripped from our beam and brought down there. The place seemed staged—it was clean and quiet, the tooling machines stood at ready, and we looked out through a clear plastic barrier into the big dark space of the museum's floor. Night view from the fish tank, I thought, as Harry looked bashful.

We found Ned at his enormous toolbox. One doesn't actually greet Ned, so much as simply move into his presence, where he notices you. A tall lanky guy, with a dark head of curly hair, he gathered us in with an unsurprised gaze. The big toolbox was brown, had wheels, was as tall as a refrigerator, and was adorned with two wooden dragons. Ned was locating a certain wrench among an array of wrenches in one of its steel

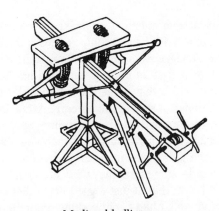

Medieval ballista

From *Dictionary of the Middle Ages*, Joseph R. Strayer, ed.,
Charles Scribner's Sons, New York, 1983. Copyright © 1983
by the American Council of Learned Societies

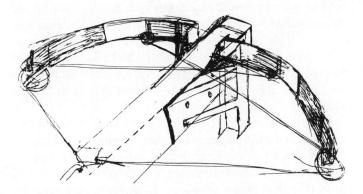

One of Harry's design sketches

drawers, but when he saw us, he put the wrench back in the drawer.

Ned was at that moment engaged in making a weather model, one of a series he has composed. Out there in the dark beyond the Plexiglas was his tornado model, a big booth in which a fog vortex spun when you turned on the fans. In its base was a machine that broke water into molecules by ultrasound, and this mist rose in the air current, making the vortex visible. The column twisted like a spinning rope, or undulated, transparent and tubular, or burst into uncoiling ribbons of mist. Sometimes the kids at the museum slashed at Ned's tornado with their hands; sometimes they tried sitting in the booth with it, which killed it; then it just fogged up in there.

That evening Ned had been working on preliminary experiments for a global weather model. Ultimately, he would fill a glass ball with an iridescent blue-and-white fluid, and stir its contents with a small, quick propeller spinning at the center of the globe. Beneath the glass surface, the liquid would turn and separate into blue and white, revealing its currents in spirals and waves. The patterns formed in bands, like the clouds of Jupiter. Ned had begun the work by swirling this fluid in ordinary bowls, and observing it closely. Once he had demonstrated this technique for a group of schoolchildren, who sat in bewildered silence as Ned simply stared into his bowl at the blue-and-white soup. Finally he noticed them and tried to explain. "Have you ever seen those satellite weather maps on TV?" he asked them. "Oh, is *this* how they do that?" one said.

But for that evening Ned had agreed to help us with our catapult, and so he put his globe aside. When I had first spoken to Ned about the catapult project, he had expressed an uneasy interest in it. I'd actually thought he disapproved at first, until he offered to let us use the shop. His was a singular response to a project that most people either loved in a little-kid way or said sounded weird and dumb. That night I realized that Ned felt both of these responses, and this reticent enthusiasm made me warier than any unalloyed response, as if there was indeed something subtly dangerous about the project, some harm less obvious than the danger of, say, being impaled by truck spring shards.

We laid the parts of our stone-throwing mechanism on Ned's neat workbench, where they looked, as Harry had insisted, rather crude. It was getting through my head, I thought. Ned looked over these parts rather carefully, as an anthropologist might. He picked up one of the triangular plates and asked us what it was for. I clanked the thing into the angle between the wing mount and its baseplate and said the word "gusset." Ned got it. "Maybe my friend Tom will help with the welding," he said.

Next Ned picked up the trigger pin, the J-bolt we'd purchased instead of the fireplace poker. Harry explained the trigger to Ned. We would hinge that pin on the beam's steel saddle, like those birds that dip for water. Harry thought that we needed to modify the pin a little, to chop away some of the inside tip of its crook. That way we'd lessen the trigger's angle of contact with the bowstring—taking it back from the perpendicular, Harry said—encouraging the string to pop out of there. Ned got that, too. We could use the band saw, he said.

We could have drilled a hole through the pin to hang it on its hinge, but we thought that the hole would weaken the steel, so we had decided to weld a tab onto the pin, and drill the hole through that. We had decided to do that, anyway, until we happened to find a metal pulley wheel at Feeney's with its center hole already the right size. Since we didn't have our own drill press, we thought we'd just weld this pulley wheel onto the pin, and save a step. And of all the stuff that we had dumped on Ned's workbench, this pulley wheel held his attention the longest. "All we have to do with that," I said, "is chop one side flat and weld it on there."

Ned was scratching thoughtfully at the pulley wheel with his thumbnail. "Right," he said. "Do you know what this is?" he asked at last.

"It's a little wheel for a pulley," I said.

"No," said Ned. "I mean do you know what this is made of?"

"Steel?" I ventured.

"I don't think so," he said. Then he turned around and yelled "Tom!" into the rear of the shop area. After a moment this young guy, his hair bound into a ponytail by a bandana, came out from behind some kind of metal barricade. He wore leather welding chaps and gloves and had his goggles pushed back on his forehead. Into Guns n' Roses, I thought. Ned showed him the pulley wheel and asked, "Would you put a torch on this?"

Tom took the thing in his blackened glove, looked it over and hefted it. "Probably not," he said. "Did you try a magnet on it?" Ned got a magnet from his toolbox, and Harry and I stood there like parents among surgeons. Tom put the wheel on the workbench and Ned prodded it. The pulley clicked uselessly away from the magnet—there was no attraction.

"Pot metal," said Harry.

"It's not steel?" I said.

"It's some kind of cheap, light alloy," said Ned, "like tin and zinc. Sometimes there's even plastic in it."

"What happens if you try to weld it?" I said.

"It lights on fire and gives off toxic fumes," said Ned.

"What did you want that pulley for, anyway?" asked Tom.

"We didn't really need it," Harry said. "The hole in the middle was the right size, that's all."

"Yeah," I added. "What this is is a ready-made hole."

At that point the four of us stood around the workbench, not saying anything for a second, but having a sort of meaningful moment. Ned and Tom seemed to exchange glances, and I could detect just the barest hint of irony in the air. Finally, Ned broke the silence. "We'll make another one out of steel," he said. "This stuff should be easy. Tom and I will even do it for you."

"Yeah," said Tom. "We'll do this stuff for you."

So at the science shop Harry and I never actually touched the machines. We stood around and offered our opinions and watched Ned and Tom do the work. More particularly, I stood around watching Ned, while Harry went back into the welding area with Tom and stood around watching him work back there. Ned trimmed our trigger tip with the band saw, a machine that raced to a high-pitched whirr when he turned it on. When Ned lowered his goggles and began to cut, the band of the blade poured out of the saw's housing through a stream of milky fluid and into our piece of steel, through which it exited into the saw's big base. When he was done he gave me back both pieces, the trigger pin and the slim columnar wedge he'd trimmed off of it.

After a while, Harry came out of the welding area. "That guy's pretty good," he said. "He fixed up our welds." He added, "We found the robots." I followed Harry back into the burnt-steel smell behind the barrier. Our wing mounts lay there on the floor by the canisters, cooling. I looked them over. In all their interstices was new welding, smooth straight beads where our spatters had been. Plus they'd been gussetted.

And in a corner of the welding shop lay Chico's robots, like skeletons on the plains of Africa. There were two of them, in a tangle of hydraulic tubing. One was made to pick up a rock and throw it; the other simply to rock to its feet. Either action took a lot of work, a lot of manipulation, Tom said. But the crowd always cheered them on like crazy, he added. Ned observed that these things seemed truly emotional as they flung themselves around in their violent, robotic fits.

I had invited Ned to go to dinner with Harry and me, and when they finished the work, we waited for him to put his tools away. I knew that Ned was a deliberate and observant man; I had seen him watch fog, or the movement of a crowd at a distance. I'd been with him in his car, and I had seen him choose his route by what he felt like seeing on that particular day. Even with dinner waiting at the end of his workday, he put his tools away in this same spirit, looking at each drawer in the big toolbox before he closed it, pushing them quietly shut with one foot as he stood firmly balanced on the other. As he did it he said, "I'm putting you away now, tools."

We walked under streetlights through the Marina to the restaurant. Ned said that he had made a catapult once, in college—one with a big bucket and not much throw that dumped water out of the dorm window onto football players. This was just right for Ned, I thought, a big gusher, not arcing meanly through the air, but effective, a drenching downpour, laid on with the accuracy great volume assures. We walked to a place on Chestnut Street which has a voluminous menu describing about a hundred kinds of sandwiches, each of which is named for a former patron. Harry figured out what he wanted—tongue—and asked the waitress which sandwich it was. Ned read through the list. I got what I always got. I asked Ned if he wanted to come out to the Headlands on Sunday and watch us fire the catapult. Maybe he would, he said.

CHAPTER 22

The Setup

Saturday morning, the day before the firing, Harry and I bought gloves. The tool rental place in Berkeley had a small selection of merchandise designed to appeal to people renting tools, including a rack of gloves, which appealed to me, anyway. They had plenty of stock on the rack, dozens of pairs, in several kinds: gardeners' gloves, white with nubbly rubber palms; canvas gloves for painters; and blue denim with stiff cuffs, railroad gloves. I began trying them on. I held up my hands in the denim ones for Harry, and said, "Check it out—John Henry hands."

Harry couldn't have cared less about gloves. He had thick palms. He worked with sandpaper a lot, and varnish, which hardened into his whorls. Harry could pick up very hot things and not be hurt. Once he handed me a plate at the table, and I couldn't take it out of his hand. Plus a nick or a cut didn't bother him much. Sometimes I even had to point out to him that he was bleeding, always astonished at his indifference, even after he noticed. Harry thought about this differently, of course. "You're afraid of pain," he said to me. That's why I couldn't get loose on a skateboard, he said. It was true: I was afraid of hurting my hands, which you usually do, falling off a skateboard. I had to admit I was afraid of pain; on the other hand, maybe it hurt me more. I hated to hurt my hands. I hated to have anything on my hands. I have broad hands, like paddles, but they stay indoors mostly, and usually don't pick up things like steel. So perhaps I was a little thin-skinned, literally.

In any case, I wanted the gloves. I did not want, however, to be the only one wearing them at the shoot—as we'd come to call the firing of the catapult. So I offered to buy gloves for us both. "Come on, Harry," I said. "Look, we'll be hauling around rocks; and that beam has serious splinters, besides."

"The steel," said Harry. "That's the thing. The stuff does wear on your hands." He picked up a pair of thick leather gloves. "With steel one false move and you're *ffftt!*" he added. He snapped his fingers as he made the sound.

I picked up another pair of the same kind, and we tried them on. They were nice, soft, yellowish split pigskin—the toughest hide you can get, Harry said. They had a short nap like suede, three neat seams down the back of the hand, and reinforced thumbs. We both liked them. We paid for the gloves when we rented the comealong, and pulled the tags off them as soon as they were ours. We stopped in the parking lot to put them on, stretching our fingers inside them and punching our palms to break them in.

Harry wouldn't even drive in his gloves, though. He took them off when he started the truck, saying that they interfered with his contact. Wasn't that the whole point? I said. I liked my gloves, and wore them for the rest of the day. But I'd bought them rather too large, and they'd make me look ham-handed, apelike, in the pictures from the shoot.

We got pictures because I had called my friend Howie earlier that week to ask him to photograph us firing a catapult. Howie, a wiseguy, said he'd do it as long as we weren't planning to shoot anything back at him. Harry hadn't wanted any witnesses at the shoot, period. I'd wanted to send out some invitations, but he had flatly refused—to the point of threatening to quit—if I did any such thing. Nobody could be there, he said. He argued that it was way too dangerous, adding, "Besides, it might not work."

So at first, he had even objected to having the catapult photographed. He'd said it was less pure—that the best thing would have been to have simply fired it ourselves in the dawn, mission accomplished, end of project. But for one thing, I had argued, we never would have done it if it had to be done so purely. And for another, we would need some pictures to show to our lecture audience. If we didn't have them, I said, we'd have to talk for the whole hour. Confronted with that wasteland, Harry had agreed to let Howie take some pictures. As a precondition, though, he made me admit that the pictures weren't just for our audience. I had to say I wanted the camera there myself. And I had to say it sincerely. So I called Howie, and got him started on the assignment by coming over to my house, and taking some pictures of our little model and the Red Creek quartzite. The photo of the rock was merely symbolic—I'd become attached to the ancient pink stone on my hall table, and had decided not to use it for ammunition. We'd just shoot the granite. A rock was a rock, I thought, wanting it both ways.

About that time the first press notices of our catapult lecture were beginning to be published. An editor at a local magazine had taken it upon herself to embellish the art center's press release. She'd written that Harry and I had our "dander up." "They're not exactly at peace with the universe," she said. We'd made "Warrior Art," she said. Harry was disgusted when I showed it to him. "Oh, brother," he said. "That's why I never go to openings."

"Warrior Art" made me uneasy. Plus I was feeling some guilt about lying to our sponsors. I'd called JD at the art center, and told her that I had to tell her something about the catapult. What? she'd asked.

"It's just serious, that's all, it's a serious weapon." JD and

I had joked about the catapult for months now. All along she had thought the project was cute—weird and male, but cute. "How serious?" she said.

"Real serious. Not so funny anymore. Big."

Then she'd reminded me of our pledge to use mock rocks. "Right," I'd said, which might have been taken to mean that we were indeed planning to shoot mock rocks, rather than the grapefruit-sized granite fungoes we hoped to be heaving. Mercifully, she didn't ask me any more about them. She just reminded me to bring my copy of the permit, in case the rangers came by to check on us, and we hung up. I told Harry about it, hoping he would console me. "Don't worry," Harry said. "We'll get out there really early, and fire the thing before anyone gets up there." I could tell the situation was playing into Harry's wish to have no audience, but I took the consolation, anyway. The prospect that we wouldn't get caught made me feel better about lying.

We got back to the porch at Harry's about noon, and unloaded the truck—me in my gloves. On the porch we dropped the rented comealong, a big, battered sucker, some other hardware we'd bought and our new bowstring—the cable net—which Rick had proudly presented to us at Feeney's. Harry's baby, Julia, was sitting on her swing on the porch when we got there. The porch was covered, and Susan had hung the swing from the roof beam. When I brought in the cable net, Julia looked up and said, "Swing?" The cable net looked like a woven playground swing, at that, the seat where the stone would go. "No," I told her. "You have a swing. I have part of the catapult."

"Cat-pull," she said.

On the porch, the beam looked different, better. Harry had carved a bowled trough into the wood between the spring mounts, so that the stones would have a little more clearance in there. Inside, the fir was quite golden where he had relieved

it. Harry had also cut a maple plug, about the size and proportions of a stick of butter, and had drilled a hole through
one end of it. As we talked about how we should proceed, he
proceeded to finish this part. He unbraided the end of a length
of Manila rope, drew one of its ends through the hole in the
plug, then rebraided the strand into the rope, rolling it under
the sole of his foot, finally, to mesh the braids. I was impressed.
"Where'd you learn to do that?" I asked.

"The guy in my father's shop, Red, he taught me. Red liked
me," Harry added. "My father was jealous."

"You really like knots, then?" I asked.

"Never liked knots," Harry said, twisting the rope.

Next we hooked up the new stuff, installing the monster
comealong and stringing the bow for the first time. Stringing
the bow was a rigidly logical two-step process: We hooked
chain to bow and drew it taut, pulled back far enough so that
the bow's arc would accommodate the bowstring. "Just a
gentle arc," said Harry. Then we worked the loops at the ends
of the bowstring through the notches on the bow, and backed
the chain off until the bowstring took the tension and the chain
could be removed.

We had used two undersized C-clamps, picture framing
clamps, to link the chain to the bow—clamping them just
ahead of the bow-notches—and these guys didn't look so
happy about that work. They were rotten clamps, and they
bent a little more each time we strained them. They'd manage,
we declared, though we both continued to eye them as a source
of potential disaster. We strung the bow anyway, and afterward, cleared of the installation hardware, it was tight. The
cable sung bassily when I plucked it. Very cool, I thought.
Harry just looked carefully at everything. We left it strung as
we worked on the rest.

Next we screwed the trigger mount into the beam, first
drilling three sets of screw holes, so that we'd have three power-
settings for shooting stones. The furthest setting seemed ri-
diculously far back on the beam, Harry said. To draw the bow

that far back would be to bend the springs into a full arc. "A one-eighty," he said, making that arc of the drawn bow with his arms.

Julia came over as I was putting in these holes, and pretended to help me. She picked up a loose nut, set it on the beam and rotated it, saying something over and over. "I want queen teaser bed," I heard, at last. When I finished drilling, I said, "You want queen teaser bed?" "No," she said, speaking very clearly now, "I want cream cheese on bread."

After that Harry and I realized that we could set the whole thing, for the first time drawing the bow and engaging it in the trigger. We hooked up the comealong—this time to the bowstring itself, and I jacked it back, quickly at first, anxiously at the end, until the crook of the trigger pin might catch it. Harry dropped the crook over the cable, and jammed the wooden plug behind the pin's fulcrum. Then we backed off the comealong and removed it. The trigger held, and the bow sat there cocked, pure potential. All we would have needed to do to fire it would be to load up a rock and yank on the rope, pulling the plug out of the trigger. Pow. Harry said it would go right through the wooden wall of the porch, probably. Still, he couldn't relax and enjoy looking at it, though. He was worrying about how we were going to load the rock, examining the taut net, fretting that the basket was closing shut under the strain. It could be grabbing the rock, he said. After that, we hooked the comealong back up, released the tension, and unstrung the thing.

Harry's big white house had been divided up into apartments at one time, and the former owners had built an exterior staircase to the second floor for the upper tenants. There were no upper tenants now; the house was opened up, and this staircase wasn't necessary. So Harry tore it off the house. He worked like a madman, said Susan, reducing the staircase into a pile of whitewashed timbers in a couple of hours. The pile lay by the house untouched for months, until we needed our last catapult part, a stand for the bow, a kind of tripod that would point the beam upward in the proper angle for trajectory.

As we rooted through the pile for decent four-by-fours, the two boys came out and began to horse around on the porch, yelling and kick-boxing the air. "Hey, be quiet, if you're going to be out here," Harry said. So Isaac and Ross decided to make mazes for their animals—a lizard named Frankie and a yet-to-be-purchased frog. They dragged some of the scrap lumber out of the staircase pile, and took over the other half of the deck, pounding with hammers and shouting for help now and again, as Harry and I sawed, drilled and hammered on the catapult stand. To hold the four-by-fours together, we toe-nailed an upright into the middle of a base, and reinforced it with screws through tie-strips of thin steel plating, called Simpson's Strong-Tie. Then we braced the upright with diagonals until we had a big bisected triangle of timbers. We could pin the point of the triangle under the snout of the bow, and have a stable base supporting it.

For a sixteen-inch gun, I remembered from my weapons reading, forty-five degrees was the trajectory's angle for maximum range—higher meant shorter. We made ours a little steeper than that, and so our ten-foot stock required a stand about seven feet high. By the time we'd finished the stand, what daylight we had was failing. Harry went in and turned on the porch lights, so we could put the whole thing together.

But by then the whole thing was too big for the porch. The catapult seemed to have grown larger and larger before our eyes. Harry remembered, looking at these enormous parts, that he had initially imagined using a five-foot board for the stock. Now the whole thing had become so huge that we had to drag the stuff out into the backyard if we were going to assemble it. I'd been hauling things around all day, and when we stooped to pick up our new stand, even that seemed stupidly heavy. We half-dragged it out the screen door into the small backyard, and leaned it against the fence. There was a little light left out there—just some high clouds glowing in the west beyond the bay. We stood out there a minute, Harry wiping his hands on his jeans, me in my gloves. Then we went back in and tried to lift the bow assembly. At first I didn't think I could move it. But at Harry's urging, we each got under a wing, and pulled the iron-clad roof beam down the back steps, its base thudding three times behind us. We lay it next to the stand. I sat on it.

We still had to lift the nose of the stock and set the bow on its stand, on a bolt that would hold it in place. After a moment's rest, Harry said, "Let's do it."

At first, I'd thought that one of us would hold up the stand, which wouldn't be stable without weight on it, while the other put the bow on it. Now that I'd felt the weight of the thing, I knew that was out. Both of us would have to be under the weapon. So we balanced the stand on its four-inch width in the middle of the yard and grabbed the wings again. We braced like weight lifters and at a signal boosted the steel over our heads. Holding it up there, we stalked toward the stand, dragging the end of the beam behind us, the stand glimmering whitely in the dim yard, and seeming a long way off. I couldn't keep my wing up as we approached it, and was just a fraction of an inch too low when we got there. I rammed the stand with the stock, knocking it over and making it slam into the ground. Cursing and groaning, we lowered the steel. This time we both sat on the stock.

By the time we tried to hoist the stock again, we were punchy, and the whole business began to strike us as funny, all this effort for a catapult. Just as we boosted the wings again, we started giggling and had to drop the thing, hopping out of the way of the heavy beam as it hit the ground and we choked with laughter.

After that we got grim. It had to be done. No laughing, Harry said. And this time, crying out with the effort, we finally put the bow on the stock. There it stood, dwarfing us, aimed inadvertently at the only lighted window in the neighboring house. We moved back to look. My shoulders ached; my back hurt; my arms felt bowed and stiff; my gloves hung from my hands. But when we saw the catapult there in the dimness, bearing its arched arms aloft, we were elated. It was quite a thing. "Yes!" I exclaimed, as Harry made heavy-metal guitar noises, and went around viewing the catapult from different angles. It was bigger than I thought. It was dark and cruciform, the steel plate, the shiny hex-bolt heads, the springs giving it a mean and purposeful look. "Did we build a weapon or what?" I said. Harry went into the house to get Susan.

Susan was not in the mood to admire our weapon. She came out onto the dark porch, and stood in the light from the door. "I don't like it anymore," she said, wiping her hands. "It's not funny—it's cruel. Take it down before the neighbors see it, or they'll think there's some insane survivalist living on the block." This said, she turned and went back inside.

Harry and I stood around looking at the thing for a while in the dark. Then Harry said, "You think it looks cruel?" and without waiting for me, answered himself. "She's right. It does look cruel."

I hadn't taken the catapult seriously for a while. Maybe I never had, really, I thought. Even at first the project had been wholly funny, a fun thing to do. I'd begun a list of places to lay siege to—The Bank of America, Stanford University— cheered by the fantasy of attacking something as complex and civilized as an institution with something as dumb and brutal as stones. Then, when we got the money—or actually so that we could get the money—I had had to try to take the project seriously for a time, to imagine that we might make this weapon to demonstrate the weapon-maker's mind. Even then, though, I hadn't considered that we might actually assume that mind. We would just put on the mask, as it were.

But stronger forces had taken over—or rather, had been there all along—and these had reduced my reasons to alibis. I'd fallen for some natural attractiveness about the weapon, the attractiveness of engineering—of getting the world to do what you want it to—and the attractiveness of destructive power, harder to acknowledge and more powerful for that.

So that last night I hadn't really had a thought for weeks. The weapon had possessed us, I thought, as we stood in the dark. And Susan was right—the neighbors would see no irony in it. There we were—it was us, Harry and Jim, not a pair of gun freaks—amidst this quiet neighborhood of families, shouldering this big brutal object, a thing almost too heavy to bear, and pointing it at the house next door.

But this thought, too, was a passing reflection. Even seeing it like that, whole and from the outside, wasn't enough. For in a moment we were the catapult's again, living and dying to shoot the rocks the next day. With our last effort we took it down and dismantled it, then backed the truck through the gate into the yard and loaded the parts, sorting materials and tools, packing a kit of emergency parts, wedging the heavy end of the stock against the tailgate and tying the beam over the cab. Nobody would guess we were transporting a weapon, said Harry.

"Sure," I said. "We can always claim to be serious fishermen." At the last, Harry went down to his basement studio, and brought up our cardboard box full of the chunks of Yosemite granite. We snugged it into the truck bed, and quit for the night.

I went home, weaving in the bridge lane with my need to sleep. At home I fed the cat, set the clock for five, and collapsed. When the clock went off, it was still dark.

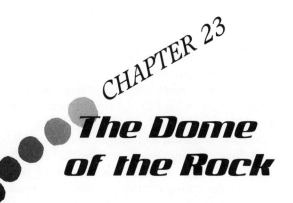

CHAPTER 23

The Dome of the Rock

About a year after Harry and I fired our catapult, I found myself in Jerusalem, still thinking about catapults and sieges. I had exchanged day for night on the plane, had arrived at Tel Aviv in the dark, and had taken a stretch limo called a *sherut* upslope into Jerusalem. I checked into the Y on King David Street, ate some gorp, and went to bed wide awake.

So the sun rose without the old meaning for me, since it was just getting dark back home. I was still awake, and with nothing to do so early, I got a paper and went out to find some stones. Here as elsewhere, I spent my spare time picking up rocks. In the Valley of Hell, the original low place of that name between the Y and the Old City, I wandered through a construction site, where amidst the cold machines and piles of rubble, a bullet-pocked office building was going down, a shopping mall going up. I looked around in the rubble: chunks of limestone and disintegrating chalk, mostly, among squared pieces of cement sprouting tangles of steel, and, pink among the pale heaps, some fragments of broken terra-cotta, roof tiles or pots maybe, maybe old.

I crossed the road and climbed the hill beneath the walls of the Old City. The road too was being worked on: a new ramp replacing the old viaduct, and revealed in the earth beneath,

Babylonian-age ruins, ruins which would be reburied after the construction, the paper said. They were merely ordinary Babylonian ruins, it seemed. I looked at these old stones, and remembered about being from America, a new country on the unbelievably ancient earth.

But here in Jerusalem, too, was the old earth, the bedrock marine sediment, limestone. Limestone might not be called stone at all, I thought, because it is not directly the body of the earth but the bodies of creatures who lived upon it. The plateau of marine sediment upholding Jerusalem is a thick layer of ancient bones, fossilized, the bones of creatures who died in a sea where Jerusalem came to be, a sea no person ever saw.

And this limestone didn't just lie beneath the city—it *was* the city, always had been. The stones from the Babylonian time were cut limestone, as was the high, variegated wall of the Old City, which rose before me, its blocks cut in the varying styles of the conquering empires. Huge Herodian blocks pushed up through the steep earth at the base of the wall. Atop those lay the stones of the Turks; and upon those, the stones of the Crusaders.

The more recent, the more rude, seemed to be the rule. The Crusader stones were most irregular, some seeming simply mortared and piled. But Herod's stones were perfect. Each fit, planar and perpendicular with the next; each had an inset border—to sluice the rain off, maybe, or simply to impress.

At the top of the hill, I walked through the Jaffa Gate, and wound into the Old City's interior, a maze of vaulted limestone. On a side street I saw two Arab stonecutters working, cutting stone as they sat on the stone pavement, their chips of broken rock heaped around them. These men were cutting the stone into blocks with hammers and chisels, working by some ancient technique—the chisel in the right fist, its tip held between the ring finger and the pinkie. The Stonecutter's Grip. Their tools moved casually and exactly—there was no wasted force behind the hammer; the chisel found its precise way. Their new stones and the really old stones, their chips and the pavement they sat on were all the same—all limestone, whitened with chalk dust, like Herod's impressive stones, or like the stones of Golgotha, for that matter, the slabs beneath the censers and silver icons. The whole Old City was the same pale limestone, blank enough to seem to change tint with the time of day. The wall that looked gray that morning would be pink at dusk.

So I ended up walking around the city all day, looking at limestone worked everywhere into the old human scheme of things. Beneath the Dome of the Rock, where God is said to have spared Isaac and where Mohammad is said to have ascended to heaven, azure and gold filigree enshrines the same marine stone. At one corner of this Persian shrine, the guard bid me put my hand—my right hand, please—through an opening in the marble altar, to touch the stone. In there, millions of palms had polished the limestone almost as smooth as marble.

Later I went with an Israeli guide to the Western Wall. He pointed out the former ground level on the wall, about ten feet above us. He said you could tell where it was because of the marks above it—places the ancient Jews had worn it down, he said, beating their foreheads into the limestone for ages of grief.

There was a siege on at the moment, as usual in Jerusalem, this one mostly a matter of kids throwing stones at soldiers and soldiers shooting tear gas and rubber bullets back. Just that day, in fact, some Palestinian children would throw the pieces of limestone that lay in their schoolyard, first into the commuter traffic, and then at the Israeli soldiers who showed up to restore order. The young soldiers had to fight a pitched battle to reclaim the schoolyard, finally subduing and jailing a few of the stone-throwing children.

For their part, the children had been protesting a parade. A group of fundamentalist Jews had paraded in ancient garb through the Jewish Quarter, calling on the government to seize the Dome of the Rock, the site of the Holy of Holies. It was time, said the leader of the group, to rebuild the Third Temple, on the site of Solomon's and Herod's temples. And it was time, he said, to destroy the abomination called the Dome of the Rock that stood upon that place now. As part of their procession, a flat-bed truck bore a three-ton cube of limestone, which the group proposed to lay as the first stone of the new temple. Police kept the group away from the Dome of the Rock, and moderates in the government labeled them lunatics, but in Beirut the Hesbollah took their threat to the Dome of the Rock seriously and urged the schoolchildren to show their resolve to hold the Temple Mount, the last outpost of Islamic control in the Old City. The children responded with what came to hand. A year later, the same situation would end with more stones and the killing of eighteen Arabs on the Temple Mount.

Rock-throwing is an ancient tradition in this place. A piece of this same limestone David probably flung at the giant Goliath, when the latter's size embodied the dominance of his tribe, the Philistines, in that land. That was some thousands of years ago, before Greek invaders had translated "Philistine" into "Palestinian."

By the time I left Israel, I had again gathered quite a few rocks, different rocks, though, from those I had gathered on trips before. These Holy Land rocks I valued for the evidence that someone had worked them. I was thinking in terms of history, at that point, not geology. My luggage had grown heavy with that limestone, shaped in one way or another by human hands. I had collected, among other pieces, rubble from Hell and a chip from the stonecutters' pile. And at the end of the trip, I had spent a few days looking for rocks in the desert, as with a guided group from the Society for the Protection of Nature in Israel, I climbed mountain trails around Masada

Woodcut by H. Clewans. Copyright © 1990

and explored springs in the rift wall above the Dead Sea. In some places, the guide said, ancient water finally rose to the surface here, water a million years underground, fossil water.

There in the desert, I found even more rocks, some really old ones, at least in human terms. On the peak of a West Bank hill, near a lookout with new trenches and gun emplacements, I wandered around in the ruins of a Chalcolithic temple, six thousand years old, as old as the story of Adam and Eve. Then, at the end of my trip, the SPNI bus left me at Ben Gurion Airport near Tel Aviv. Unbathed and unshaven, I looked like what I was: someone who had been in the wilderness for several days. Naturally I got searched. I was taken to a small windowless room with a small table, no chair, where an agent, a kid of about twenty with a machine gun, asked me questions—

the first in Hebrew—and made me empty my pockets, then took my luggage thoroughly apart. When he got to the rocks, he pulled them out one at a time and set them in a line on the table between us.

I said, "Souvenirs."

"You don't have to tell me," he said. "I always bring home the stones."

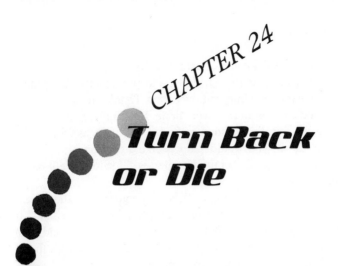

CHAPTER 24

Turn Back
or Die

It was worse than dark when I left my apartment that morning—a thick fog further obscured things in the street. I saw the fog swirling in my headlight beams and swore. Even if we could shoot stones with our catapult, we'd never see them, at this rate. The freeway standards swam by, their yellow cones thick with swirl. Please let it lift, I said to no one, to the god of fog.

I plunged onto the lower deck of the Bay Bridge, and rocketed along in there, nobody around me, only a single pair of brake lights up ahead in the mist. It would be clearer in Oakland, I hoped. Please, I thought, some light. I shot through Treasure Island, over the cantilever section, and down the ramp on the far side. And when I emerged and could look skyward again, it did seem barely brighter. But then it usually was in Oakland. A few miles east, I griped, and there'd be clear skies, a desert sun over Livermore.

At Harry's house I peered through the fence into the backyard first of all. The truck sat packed and ready, nestled tail-in along the side of the house, the beam projected over its cab like a lance. When I knocked, Harry let me in. He was up and dressed in jeans and a black Industrial Light and Magic sweatshirt, but he hadn't finished breakfast yet, he said, so I followed him back to the kitchen. Susan was up, too, in her robe and eating Cheerios. Harry sat down and put sugar in his coffee. He was eating stuffing, left over from Thanksgiving.

I stood there until they were done. Mercifully, nobody mentioned the fog. But as soon as Susan had seen us off into the gloomy dawn, Harry said, "Fog," like a caption.

"We'll be high up on the Headlands," I said. "We'll be up over it."

"Up over it?" Harry said. "The fog comes from the ocean. The Headlands are on the ocean." Mere hope never made it with Harry, I thought. But as usual his pessimism made me feel better. It made me face things, if only to argue with them. Besides, Harry was sometimes wrong.

"Maybe it'll burn off," I said.

We checked the tie-downs and the parts. Everything was still fine, the knots still tight. So we got in the truck, and Harry drove out of his sleeping neighborhood, picked up the freeway and headed north along the bay toward Richmond. We'd decided to take the Richmond Bridge, from the East Bay to Marin. Going the other way, we'd have to cross two bridges and negotiate the city with a catapult strapped to the truck. But in Richmond, deeper fog clouded the freeway. I whined in the shotgun seat.

The Richmond Bridge is towering, grated, see-through, and as we climbed to its peak, I strained to see some greater clarity above. And at the crest, I could. We broke through the surface of the fog, and there was sunlight on the superstructure. There we looked for a second across a landscape of fog, Mt. Tam like a big rock in rough surf on the far side. "Seems like it might clear," I said.

"Marin," said Harry happily. "Land of Enchantment."

The Richmond Bridge makes landfall at San Quentin. There at the shore we passed marshes and a Rod and Gun Club. Probably for the guards on their day off, Harry said. I saw a mallard drop into the cattails. It was definitely less foggy, I thought. But as we headed south down 101 into Mill Valley, we got socked in again. Harry surprised me by saying, "There's usually fog in the valley," and he was right, because as we again began to climb out of it up the Waldo Grade until we emerged from the tunnel at the top, we found clear daylight across the Golden Gate. The strait sheened below us, and far across it, the big tripod antenna stood crisply atop Twin Peaks, striped red and white. We were going to be able to see, I declared.

We drove south almost to the Golden Gate Bridge, took the last exit and climbed into the Headlands. The steep road verged the cliffside above open ocean, the Pacific gathered there by the wide strait and funneled down to the merely monumental scale of the bridge, which fell behind us and away, looking more and more like a delicate toy, the city and the bay gathering still smaller behind it. I looked back, and informed Harry that we could see all the way to Oakland. A perfect day for a siege.

We climbed past the biggest bunkers, the World War I model called Kirby where Arnold Palmer hit his drive, then at the crest past the elaborate bunker system called 129, built for World War II. From there the road narrowed, became one-way only, and plunged out and across the cliff face, beyond which Point Bonita coiled below us like a fist on the water. The road picked up the crest of the spine below, and about a half-mile farther we came to Battery Rathbone, our bunker.

Rathbone was a bunker that protected the big bunkers, an outpost where a smaller gun might be directed against aircraft. The gun was gone, of course, and the ruined bunker was just a concrete bowl—a turret base—with an apron and a lookout, installed there at the crest of the ridge. Below were ammo storage cells; above, the pillbox for the lookout.

"Good," said Harry, as we pulled up. "Nobody's around."

We got out of the truck in our gloves, our jackets zipped up.

We untied the beam and carried it on our shoulders up the concrete steps to the lip of the bunker, where we set it on the gravelly apron, and stood a moment looking off the cliff. The slope fell radically, then vertically away below us. The city's western avenues striped the shore on the far side of the strait; the ocean broadened for some miles and disappeared into the retreating fog.

"Let's point it at the Avenues," I said. "The ocean's too big."

"I think the Avenues might be out of range," said Harry, deadpan.

After that we went to work, not saying much. We made several trips each to the truck, and got out of step with each other—Harry just dropped his loads—so that we began to meet on the steps, one of us empty-handed. On his way up the steps with one set of springs, Harry just said "Avenues," in an avid and silly way. On my way up with the box of stones, I said, "Catapult ammo, sir!"

We laid everything out on the apron, then put the parts together speedily, practiced at the setup by now and moving like nervous guerrillas in broad daylight. One big improvement was Harry's idea: that we put the stock on the stand before we put the steel wings on it. This insight on the order of the procedure was the one brilliant thought to come out of the previous night's agonies mounting the stand. So we put the empty beam up first, chagrined to find out how easy it was that way. What idiots, I said.

I bolted on one set of springs while Harry bolted on the other, and likewise we secured the trigger mount, each to a side. "Where's your gloves?" I said, but Harry wouldn't answer. He was tense. He was getting everything right, and he was laboring in fear. All along, Harry had felt risks in this that I hadn't felt. But the park rangers and the shattering steel weren't on his mind now. And it seemed those other worries were secondary to, and to some extent ruses for his main fear, that the catapult wouldn't work, that it would prove a flop, a limp hose.

"This is a no-trial-run, one-time-only kind of thing," he said, screwing in his bolts.

He was right about that. For one thing, if we blew our permit date, I'd have to go back to JD. Still, I had never really entertained the notion that the thing wouldn't work, and didn't entertain it then. Let him hold up the worry end of it, I thought. I was blithe.

We hooked up our rotten C-clamps, yoked the comealong through its anchor bolts, and strung the chain so we could string the bow. "Come on," Harry said to the C-clamps. "You have to work just a couple more times." And peeling their plating, the clamps took the strain again. The massive old comealong clucked as it inched back the chain. When we'd strung the bow, we stood back to appreciate it. The catapult looked at home on the site, its steel wings spread against the sky. The water glinted below, and in the broad sweep of land containing the concrete bunker and the cliffside, the catapult looked right, graceful, ready.

Harry loaded the first stone, one of the darker, smoother pieces of granite that he had gathered from the streambed in the mountains, and then we drew the bow and hooked it into the trigger. Harry wrestled with the loaded basket, yanking on the cables so that they wouldn't entrap the rock. "If this thing's not going to work," he said, "it's not going to work right here, and the whole thing's not going to work."

"It'll work, it'll work," I said.

But it didn't work. Harry took the rope in hand, walked to the end of its twenty-foot length, and yanked it. Nothing—he couldn't pull the plug out of the trigger. So he yanked it harder, then very hard, and finally the plug popped out. Still nothing happened. The bow sat there cocked and inviolate, loaded, but the trigger hadn't budged. We hopped around in anxiety, and finally Harry picked up the big hammer, went gingerly up behind the machine, and whacked the back of the trigger with it. Finally it released, but—just as Harry had feared—the basket failed. He had set the stone in it so loosely that its cables had simply undercut the rock. The stone popped stupidly out, backspinning like mad, and plopped down the slope, landing about ten feet below us in a coyote bush.

I scrambled down the steep slope and retrieved it. When I got back up, Harry was still saying "God damn it" and looking at the basket. I thought I knew what to do. I got a rock—one of mine—and over Harry's protests jammed it into the basket.

"It would've worked fine if you hadn't dinked with it," I said. That cowed him for the moment and we cocked the thing again. I took the rope and yanked it. This time the trigger released, but the rock stuck in the basket, tumbling out only in the aftershock of the shot, and falling at the foot of the catapult. Things couldn't have looked worse.

"I knew this, I knew this," said Harry, rushing to the useless machine. We had failed twice, in opposite ways. Harry was frantic; he was practically babbling. He couldn't believe we had done all this work. Why had he let me talk him into this? He had never been sure just what he was getting out of this oddball project, anyway. He had risked looking like an insane survivalist, and for what? Because the catapult might be fun, and because he liked the challenge of building a machine that worked. But this one wasn't working, and it wasn't fun.

We started to argue. I said we should apply more power, moving the trigger mount to a higher setting. We'd blast it out of the basket. But Harry completely disagreed. He said we had to fine-tune it before we could apply more power. I was too tense to waste time arguing, though, and finally just started to do the work I thought needed doing. Harry shut up and stared at me, and then began doing what he wanted to do.

He found some wire and some perforated steel plate—the Simpson's Strong-Tie—that we'd used to reinforce the angles of the stand, and fashioned a rigid backstop for the basket. I pulled out the bolts holding the trigger mount to the beam, moved the mount up to its intermediate setting, and rebolted it.

We will never know which of us was right, or if we both were. Because when we were finished and had loaded and cocked the adjusted bow, Harry popped the plug, and the stone disappeared. The bow released, something whooshed, and the stone was gone.

"Do you see it?" said Harry, shading his eyes. "Where is it?"

But I hadn't seen anything. Just the bright sky out there, the broad flecked water. The stone certainly had gone somewhere though, and far enough to lose. We set to work cocking it again, this time silent with hope. "Watch this time," said Harry.

When I pulled the rope, we saw our stone fly. It left the bow with a potent rush and soared into the bright sky over the water, instantly tiny and far off as it hung for a heart-stopping pause at the peak of its trajectory, then fell away toward the sea, finally disappearing beneath the lip of the cliff. As the stone rose, Harry shouted, "Look at that!" and as it disappeared, "Did you see that?" By then I was jumping around, fist in the air, and shouting. "We did it!" I yelled. We hugged each other—I had never hugged Harry before, and even in that moment it was odd, something we should do. After that we jumped around again, yelling, "It worked!" over and over.

It didn't just work—it worked great. We shot rock after rock, honing our new skill and watching the stones fly. We took turns clambering down the steep ridge to the lip of the cliff, while the other fired rocks overhead, so that we could see more of the trajectory and watch the stones disappear into the watery chasm below. We got better and faster at loading the machine. Harry took up the big hammer and began beating our stones into better shapes for projectiles, knocking off the edges. "Modifying the rocks," he declared, squatting in the litter, a rubble of broken Yosemite granite and tools around the base of the catapult. "Uniform balls," I ordered, which we found funny. So we proceeded from fear to delight to happy industry, our mood tempered only momentarily when the weapon was loaded and critical. After the fifth shot we moved the trigger back to its ultimate setting—high gear, we said—and Harry winced to see the bow so deeply cocked, bent into a wide half-circle, the bowstring taut as a welded bar. But our stones seemed to hum, flung from the high setting.

We were so engaged in the shooting that we didn't even notice when the ranger pulled up. I don't know when she got there—just that at one point I turned around to see her standing there on the lip of the gun turret in her round hat, her hands on the hips of her brown uniform. Our friend Howie stood there with her, two cameras on his neck.

"Harry," I said. "Harry." But Harry was busy modifying a rock, not noticing, and by this time the ranger and I were looking at each other.

"Hi," she said. She had evidently witnessed the last shot. Behind me, Harry stopped pounding on the rock. Mom was home, I thought.

The ranger gestured toward the catapult and said, "That is *so* cool."

"We have a permit," I said, too stupid to even be relieved.

"I'm sure you do," she said, not really interested. She came over to look at the catapult close up.

Howie said hi. "I couldn't find you," he said. "I had to get the rangers to help me find this place."

Behind me, Harry was talking to the ranger. "You want to see it shoot another one?" he said. Sure, she said. We wasted no time taking the opening. I loaded a rock, Harry cocked the thing, and I yanked the plug, narrating while I did, "And then you just yank the rope, like this." Then I sent that stone into the wild blue yonder. "Neat," said the ranger. As an afterthought, she pointed down the cliff and said, "Are you sure nobody's down there?"

"There's nobody down there *now*, anyway," I said. The ranger laughed and we all laughed. Then she watched us fire another rock or two, and finally got back into her unit and drove off.

"Unbelievable," said Harry. "She got into it."

Howie took pictures of us after that, moving around us and clicking as we loaded, cocked, and fired. He made wisecracks and tried to get us to pose—Harry wouldn't. Then Howie tried to get us out of the frame for a serious picture of the catapult. Then he shot pictures of the day, the strait, the bunker. He crouched in the pillbox behind us, trying for an action feeling. He took a picture of me christening the catapult Mona, with a ballpoint pen. "More like bona," Howie cracked, clicking away.

When I told Harry I wished we had a further setting, meaning more power out of the springs, Howie gestured broadly to the sea and the cliffs and joked, "You couldn't ask for a better setting than this, Jim." Harry was happy. He sang a little as he worked, a song he sings for Julia, about bringing home a baby bumblebee.

So we shot our rocks, getting down the nuances of the procedure, and finally getting good enough to look for targets. At first there wasn't much to shoot at, but we noticed to our delight a white cruise ship entering the strait. So we held our last good rock and waited for the ship to fall within our sights. We got silly when we saw it. It looked like the Love Boat. It looked like an actual duck, sitting out there. When it crossed our muzzle we hurled this last stone at it, shouting "Avast!" and "Turn back or die!" The stone rose over the pale distant hull for a moment, then fell harmlessly away toward the water, the boat still a mile out of range.

Finally, as a closing festivity, we loaded the chips and fragments and fired them as a load of grapeshot. The bits splashed into the bows and scattered around us as we shielded our eyes.

Then the cardboard box which had held our chunks of granite since summer was empty—though its sides retained the bulges where the stones had been. It was over.

Howie helped us break the thing down, taking pictures occasionally as we did, and we loaded the stuff back in the truck. I scuffed some granite fragments into the undergrowth, and the site was clear.

We took the parts inland, down the ridge behind the bunker to the art center, where we woke up the resident maintenance woman, Cinthea, by calling up to her window in the big barracks. Finally her boyfriend Tom came down and let us in. In one corner of the unused mess hall, we stowed the dismantled catapult carefully, keeping the parts with their mates. We'd have to put the catapult together one more time when we set it up in the lecture hall the following week. Harry fell silent when I reminded him of our talk.

"What are we supposed to say about this?" he said, as we emerged into the bright Sunday morning.

"We'll think of something," I said.

CHAPTER 25

Boobs and Wangers

That week I tried to think of something to say, and Harry waited to hear it. He had never pretended that he would have anything to say about the catapult, he said. That had always been my department. But I had a big problem, a word I couldn't use: Fun.

I'd had fun. I'd said we'd investigate catapult consciousness, and it turned out there was none, no war mask, no special weapon-maker's mind. I had not assumed anything; rather I'd relinquished, surrendered to unconsciousness, building and shooting the catapult. It got me. I'd lost my objectivity, if I ever had any. I'd had fun, in other words. And I didn't think we could just say that. We had fun. Thank you very much.

So that week I tried to make up some damned thing, but whatever I came up with got quashed. Harry coached me in the technicalities of the catapult. I coached him in what little I knew about its history. But it was as if the unspeakable truth— We Had Fun—displaced its pretenders. Harry was abysmal. I had a sure premonition of what would happen when he tried to repeat the details to an audience. Harry couldn't repeat them, even to me.

I was too nervous to tell Harry how badly he did—I didn't

want to scare him—and simply told him it would be fine and not to worry. Hearing that made him really nervous. "I'd better worry," he said, "because you sound like a total idiot when you try to talk about anything technical."

We were to give our lecture on Sunday afternoon at one, but by Saturday night we still had no faith in our explanations, so we decided we'd go up into the Headlands early the next day, and go for a walk, which seemed like it might help us think. We had done some thinking sometime in the Headlands, we thought. But we made a couple of mistakes, right off the bat. We dressed for the lecture, for one thing. Harry had on a collared shirt, the first one I had seen on him in a while, a pinstripe yet. I had on creased pants from the cleaners and leather shoes.

So exactly a week after the shoot, we found ourselves neat and clean and back at Battery Rathbone, where we decided to hike to the beach below. Underestimating the terrain was our second mistake. At first we proceeded straight off the ridge, the way the catapult stones had gone. Maybe there was a gap in the cliff below. But when we had skidded down to the lip of the bluff, we found it sheer—a straight drop to the beach, we thought—and had to climb back up and proceed inland along the ridge for a half-mile or so before we found an arroyo, a stream canyon splitting the seacliff, where we might descend.

At the arroyo's rim we looked down onto the tops of the small willows that lined the creekbed, and started down into the warm still air, beginning to sweat a little. There wasn't much of a path—we parted the dry scrub, and pushed down into reedier stuff, my socks picking up burrs, I could feel. Harry didn't seem to be worrying about his clothes.

Descending the dry creekbed, I remarked that the beach was a lot farther from the ridgetop than it had appeared from the bunker. It had all looked vertical, from up there. Harry was thinking the same thing, it turned out. We would have had to shoot those rocks a long way to make any kind of a splash, he said, adding, "Hundreds of yards, just to hit the water."

"Hey," I said. "Maybe we'll find some catapult stones on the beach."

The beach lay beyond the small dry lagoon into which the arroyo descended and broadened. All up and down the coast you saw this kind of estuary, like a fugal variation, I thought, complement of the cliffs of the Headlands. The stream falling down the cliff's face had worked to dissolve it, building a beach between the heads of harder rock, and finally damming its own stream and forming the lagoon. It would be marshy in wet season, though at that point it hadn't rained for months, and we crossed dry silt as we approached the sand. Without its enormous tides, San Francisco Bay would be a lagoon, too, I told Harry. It would deposit its barrier beach and silt in. Harry parried with a little of his own nature-show knowledge. Life begins in lagoons, he said.

We had to work our way along the beach in stages, skirting and climbing the feet of the cliff that stood in the surf. These Headlands seemed to become more considerable as we negotiated them. Getting around the first one, I got wet. We had to skirt between waves, and the rock face proved broader than I thought. I didn't make it. The wave caught up with me, froth and freezing ocean swirling around my ankles. Dry Harry thought it was funny. End of these shoes, I thought.

The next big outcrop stood even deeper in the surf, and there was no way around it except swimming, which we weren't about to do. But the waves had battered it nearly flat, until it was just a shelf about shoulder-high, and flat enough for us to get over, anyway. We boosted ourselves up the big rock, stepped around the tide pools and across the slick seaweed at the top, and poised to jump down onto the next stretch of beach. We stood there considering the leap, and then we saw the naked people.

* * *

The beach faced south and was enclosed by the cliff, and the naked people had set up there to sunbathe in five or six clusters on the broad sand and within sunny hollows in the seacliff. They'd set up little encampments for the day, and lay on blankets with their stuff stacked or strewn around them— towels and clothes and radios and coolers. The naked people themselves glistened and noticed us.

I was a little shocked to see them, not because they were naked—or not just because they were naked—but because I had hoped to find our catapult stones on this beach, not some sunny nudist encampment, which—now that I thought about it—I'd heard rumors of. We'd been surveying the ridge as we'd walked up the beach, trying to decide whether we had ventured far enough along the shore to reach the place stones had fallen. But here were naked people.

Harry and I jumped off the ledge together—it was a good drop and we had to roll in the sand to break our fall, as the naked people watched us. We got up slapping the sand off and tried to proceed as if normally among them, trying not to appear to stare, and not doing very well at it, I thought, looking at mesmerized Harry.

One couple lay quite nearby, and looked like models. They lay reading the Sunday paper naked, she taut-breasted and gorgeous, he broad-shouldered, both of them built, athletic, narrow-waisted, without tan lines. She had her elbows on the entertainment section, the Pink Pages, and watched us over her sunglasses as we came down the beach in our lecture clothes. "There's a path, you know," she called out.

"We know," said Harry, suddenly not looking.

The other naked encampments contained various combinations of men and women, but compared to the first pair, showing off at the top of the beach, the rest of the people on their blankets looked like marine mammals, sprawled and blubbery and happy in the sun.

At the end of that stretch of beach, a big sheer head of rock sat in the surf, and the only way beyond it was a climb harder

than I might attempt, ordinarily, especially in stupid wet leather shoes. But the naked people kept watching us, and we had walked the whole length of the beach, so we were just going to have to climb that rock as if we'd walked down there for that reason, and when we reached its broad face, we began casually crawling up it. I paused only once, when I couldn't find a decent foothold, but on my right Harry pulled himself up, his fingers in a crack, and with the eyes of the naked people burning me, I ascended after him, willy-nilly. Getting down the other side wasn't so bad, except for the drop at the end. We fell with a bump on the beach and sat there in the sand a minute afterward.

I was relieved not to have seen any catapult stones among the nudes—it would have been disconcerting to find the two together. But this beach seemed remote enough, I said to Harry. Certainly naked sunbathers weren't going to climb what we'd just climbed. So we probably hadn't bombarded naked people reading the Pink Pages, I said. Still, I couldn't help laughing about it, at my mental images of those complacent pudgy nudes panic-stricken and scattered to flight as the granite cannonballs hurled down among them. Boobs and wangers flopping, I said.

"That's not funny," Harry said, trying to stop laughing.

"You're right," I said. "It really isn't funny." Then we really broke up. We laughed so hard we couldn't get up. We lay there in the sand on our backs, screaming with laughter.

And when we could get up, we found our long-lost rocks. They weren't difficult to spot. The cliff there was basalt and chert, green and red, and the beach sand was the same, mixed to a purple-brown, on which the chunks of black and white granite, foreign stone, sparkled. And most stunningly, on a seaworn chert pedestal at the top of the beach, a big catapult stone perched like a specimen. I shouted "Look!" when I saw it. It was weird to see the stone so purposefully displayed like that, as if it was to mean something, not least that somebody had been there to display it sometime since we had shot the catapult. Or while we shot the catapult, I thought as we ran over to the stone. The display offered that obvious deduction, but was a mystery besides. Whether it was meant to show off the granite, strange and beautiful on that beach, or to display the evidence of somebody's dangerous stupidity, we could not know.

Harry took the stone off its pedestal. It was one of our bigger ones, he said. And one of the later ones, I remarked, since it showed the marks of Harry's modifications. It was still whole, though—or at least as complete as it had been when we shot it. Most of the other catapult stones we found had been shattered in the fall. We ran around the beach, gathering them up, and arguing about whether the tide had been in or out when we'd fired the catapult. If it was in, I said, we might have made a splash, after all. Harry pointed out that the tide was out at the moment, a week later. I didn't know if that meant anything.

I didn't like to think of them not splashing, but was happy to have the rocks back. We'd keep them as catapult remembrances. Plus I was glad to find no evidence that we had done any damage beyond splitting our stones. No crushed radios, victims left for dead, nothing like that.

Cradling the rocks in our arms, we lugged them back up the way we came, to the base of the outcrop where we'd landed. "I hope we can get back up that thing with these rocks," Harry said.

We looked up at the face of the big rock for a moment. It was smooth, overhanging at about ten feet, the point at which we'd jumped. A good drop, I thought. I wouldn't have jumped it at all, if the nudes hadn't been after me. Soon it became clear that we weren't going back the way we had come down— that was for sure, rocks or no rocks.

Harry suggested we try the canyon at the back of the beach, but it was a narrow beach, with a correspondingly steep stream channel behind it. We squinted up the slope, then ditched all but four of the stones, one for each hand. The canyon wasn't as steep as the rock, of course, but the footing was worse— dry dirt and gravel. "You won't fall off of it," Harry said, "but you might roll a long way."

The slope appeared to be our only way out, though, so Harry started up it with his stones. I followed, my face a couple of feet from his heels, thumping the ground with my granite as I climbed. About twenty feet up, I lost my footing and slid back a few feet, dragging the rocks like brakes. "Just come on," said Harry.

It was a long hot climb. We scrambled up the canyon, stones in hand like Neanderthals, getting more and more cautious the higher we went, until, moving very slowly, more or less on all fours, we came up under the lip at the top of the canyon. Harry boosted me up through this last gap, and handed me the stones. Then I pulled him up. Above that the ground had enough purchase to support plants, and we worked our way up the ridge, grabbing the woody shrubs. We climbed first through fennel and thistles, then into the upper scrub, through sagebrush and making our way around the colonies of poison oak on the hillside. Finally, at the ridge line, we found the roadcut and the road, and stumbled down the embankment onto the asphalt.

Battery Rathbone was just a hundred yards up the hill. We'd come up in the place where we'd refused to go down. For a long time I had been trying not to look at myself, at my appearance. There had been other matters at hand. But now we stood on the road, more or less back in civilization, and I could look, and it was as bad as I thought. I had killed my shoes; I had killed my pants. The shirt was critical. Harry was equally a mess, though not noticing. A gloppy layer of red mud lined our wet pants and shoes, and we were furred all over with dust and pollen and stickers, undercoated with sweat. I itched already.

The car clock said it was 11:55. "We have one hour," said condemned Harry. We'd be all right, I told him, making him look more miserable. I looked at the rocks that we'd hauled up the slope, and suggested we show them off in the lecture.

"That's just fabulous," said Harry. "And we can just tell them where we happened to find them, too." I saw his point. It didn't look good that so many of our stones didn't make it to the ocean.

"They don't look too mock to me," Harry added.

"Jeez, I forgot about mock rocks," I said.

Harry just looked at me, dirt on his head, and held up his rocks. "You're just going to have to tell them, Jim. We didn't use mock rocks. Big deal. What are they going to do, arrest us? So we confess. Big deal. We never have to see these people again."

"*You* never have to see them again," I said.

*   *   *

So we hid the four rocks in the trunk, finally, when we went down to face the music. At the art center I went directly to the basement, to the big Army latrine, and tried to clean myself at one sink among twenty. I washed my hands and face, and flapped at my clothes. I felt like I'd been tarred and feathered, I said.

"That would be the easy way out," said Harry. He didn't try to clean up.

"I'm not going up there looking like this," I said, picking off a few trivial burrs.

"Why bother?" he said. "We'll just get all messed up again."

"I could never get this messed up inside," I said.

"You have no choice, anyway," Harry said. He said we didn't have time—we still had to set the catapult up in the lecture hall. He was right, of course. We had to go up.

It occurred to me then that everything would have been different if Harry hadn't been there. I knew how to do a lecture; I would not have rolled in the goddamn bushes; I would have been here early; I would have been clean. I'd be sliding through this thing, telling a few jokes, showing a few slides, and sliding out the door. But Harry had to be here, I thought. And ready never to see these people again, to boot.

"I don't know how I let you talk me into this," I said, meaning into our joint appearance.

Harry got mad. "Me talk you!" he shouted. "I didn't want to do this from the beginning! This was your idea!"

"You insisted upon being here!" I shouted, our voices suddenly reverberating in the latrine.

"Just come on," said Harry.

"You did!" I said.

"Jim, let's get one thing straight," said Harry, turning on me. "I am appearing with you in public for one reason and for one reason only. I need to balance you out. If I'm up there, I won't be just a character. I'm not going to be the Igor," he added.

"Don't worry," I said, feeling trapped. "You're not Igor."

So finally I gave up on the burrs. Maybe nobody will notice, I thought. We're out here in the Headlands, in the wilds, after all. We found our parts in the mess hall where we'd left them the week before. We picked up the stock, lugged it across the big room and out into the lobby.

JD was coming down the stairs as we were coming up. She wore a black silk suit. "What happened to *you?*" she said.

"We got carried away," I said, not stopping to explain.

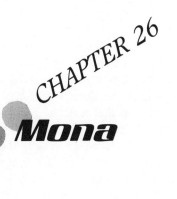

CHAPTER 26
Mona

Harry was carrying the nose, and he paused at the lecture hall's double doors.

"People," he said.

I just wanted to put the beam down. "Go," I said.

There were people in the room; it was later than we thought, I guessed. Howie was there already, set up with a projector and screen. And some of our audience had arrived: three guys with their sons. The dads had on hats with bills, and looked like their boys. The fat kid had a fat dad. They watched us bring our stock in, and when we got back from our second trip with the springs, they were already clustered around the thing, one guy squatting as if to inspect our welds. The guys didn't offer to help or anything—just pored over the parts as we delivered them.

We had more than catapult parts. I'd brought everything except my Red Creek quartzite out to the art center earlier that week—all our spare parts, the little model, even clip-on spotlights to illuminate the catapult. With all that stuff and the slides, maybe we could make it for an hour, I thought. It took several trips to just to bring it all up from the basement. On one trip, I asked one of the dads what time it was.

"Quarter to one," he said.

"We've got to get this stuff *up*," I said loudly to no one. Still, nobody offered to help. We hauled all the stuff up the stairs, and lay it on the floor, where they inspected it, then began putting it up as they watched. "Harry," I said, "maybe these guys could help us with the stand."

"You guys," said Harry, addressing the dads. "Hold this up while we put this on it." The dads obeyed, suddenly seeming more our age. Two of them steadied the stand, and the other helped Harry and me hoist the stock. "Our last stand," I joked to Harry as we lofted it. It was easy that last time, all of us under the thing like the marines at Iwo Jima.

Harry and I refused help after that, though, bolting on the wings and the trigger ourselves, hooking up the comealong, stringing the bow. This was technical, I said.

But by then the dads had their entree, and felt fine about asking all kinds of questions: what we used here, how we made that, what this was for. Harry's answer—"Truck leaf springs"—elicited their wonder. I was wondering just who these guys were. Certainly not the people I'd been expecting— the art crowd, slashed jeans and nose rings. These dads weren't opening types. From Kentfield, I supposed, maybe junior partners on their one day off. I asked one dad how he'd heard about us. A magazine, he thought. "*Warrior Art,*" he said.

Meanwhile their kids had gotten into our stuff. In our box of spare parts was the tiny model catapult Harry'd made in the summer. The boys found it in there, mangled but serviceable, and took it into a corner and began shooting paper wads or something. After a while the fat kid came over and asked if they could take it outside and shoot rocks.

"No," said Harry. "We're going to need it in here in a minute."

"It's like a prop," I explained to the dads.

Other people came in, some pausing at the door before taking their places in the pews, some coming up front to see the thing. We set the catapult up to be in plain sight from the entryway, and on second thought didn't point it there, but rotated it with the help of the dads, until it was aimed safely into the high left corner of the front. I wanted to cock and set the catapult, but Harry wouldn't hear of it, at first.

"What do you want?" he said, "Just to have the thing humming there while we talk?" After that, though, he let me cock the bow a little, just to put some definition in the springs, I said. The dads loved it. They ran their hands over the taut cable. One of them slapped the catapult on the neck, as if it were a horse. They looked disappointed when we said we weren't going to shoot it as part of our demonstration. We had slides, we said. "Gyp," said the fat kid.

The catapult looked big indoors, and so good and postmodern, I thought, surrounded by David Ireland's exposed treatment of the room, which was the big sleeping quarters in the barracks. Ireland had left raw the pressed-tin ceilings and the cast iron pillars, and these were raw with the steel of the catapult springs.

JD didn't seem to appreciate the aesthetic juxtaposition. Her face fell when she came in and saw the catapult there, and I knew she hadn't considered that the weapon would look so mean. Before she could say anything, I reminded her of my warning. "I told you it had stopped being a joke, remember?" I said.

She just said, "I'm going to let the rest of them in now," and left.

"She was mad," Harry said. Not exactly, I thought, but she was certainly something. Alarmed maybe. Maybe she just wanted to get this weapon out of here as soon as humanly possible.

Then Harry and I lowered the blinds, and Howie and I moved a ladder around, to put up and aim the spots. The rest of the crowd came in as we clipped the lights onto the overhead fixtures, and I surveyed the audience from the top of the ladder. There were enough of them to worry about, I thought, but nowhere near as many as I had expected. Mr. Standing Room Only, Harry would call me afterward.

Sara arrived, and waved from her seat next to a girlfriend. Susan had stayed home. The catapult project had begun to grate on these women, by the end. On the morning of the shoot, Sara had called Susan to suggest that the two of them raid our exclusive event, but they hadn't. They were disgusted with us, they decided. So Susan stayed home with the kids. Harry had told her it was no big deal, anyway.

At some point a group of women came in together. I was aiming spotlights—at the catapult and at the two high stools where Harry and I would sit—but I knew the women when I saw them. It was Melissa and three of her girlfriends. The Montana Mafia, I thought, nervously. Melissa and I had been writers in residence together at the Headlands, and had become good friends up there in the hills. She and her friends were like big cowgirls, good-natured but rough and tumble. When I had told her about the catapult, she'd recounted a time she'd taken her students to the beach, and had been appalled to see the little boys immediately, unconsciously, begin throwing rocks at the ducks. "Well," she'd added, "I'm just going to have to come to this thing, and see what you have to say."

That was months ago, it seemed, and I'd forgotten about it until that moment. But now here she was, I thought, and with reinforcements. They sat there in their jeans in the first two rows, all four of them, looking up at me as I looked down from the ladder. I said hi. They said hi.

When I took the ladder down, Howie shut off the houselights, putting the audience in the dark. The catapult sat in its pool of light to the left of the head of the room, and our stools sat in our pool of light, in the middle. Howie put up the first slide—his shot of our toy model, scrap wood and blue steel, next to the Red Creek quartzite on my hall table.

"After the slides," I said to Harry in the safety of the dark,

"I'll give it to you. You just say your thing and give it back to me and I'll get us out of here."

"Tell me again" said Harry.

I was annoyed. "What we said," I said. "Alexander the Greek—just how he perfected the thing—then go to the Romans—say they did it the best—and then you can just describe that picture—the Roman catapult."

"Alexander the Greek?" said Harry. "I thought it was the Great."

"Wasn't that what I said?" I said.

"Shit," said Harry, looking annoyed and out of it. Please, God, I thought.

I had gone through the slides with Howie, but I had looked at them only to admire them, and hadn't actually seen the pictures. But looking at the picture of the toy catapult with my Red Creek quartzite, I suddenly did see it, and it hit me. The rock, I thought. Right off the bat, we're unmock.

What had I been thinking? That the shot had looked great. Howie had shot the rock and the model catapult close up and below, and made them look big. On the screen the toy catapult appeared larger, thicker and more impressive than the real catapult opposite it. The darkened audience slowly grew quiet, considering the picture with its enormous-looking unmock pink rock. I had told Howie just to leave this slide up until we got started. Now it seemed we should get going, just to get the evidence off the screen.

"Go. Go." I said to Harry, nodding at the slide and adding, "Totally unmock." Harry gave me a startled, puzzled look, and then went blank as we started out into the light. I followed him out and we turned around to face the audience. Among the silhouettes was JD's—she stood at the doors, arms folded as she checked out the slide. I remembered lying to her.

I had a bad, vague premonition that the rest of the pictures might not come up to my expectations, either, and that sense began to gnaw at my confidence about the lecture, the pillar of which had been these slides. Howie had done well, I tried to tell myself, but the words, "It was all a mistake," suddenly occurred to me, for no reason I could think of, as it became clear that the audience was giving us the floor.

Harry clambered backward onto his high stool, moving as casually as he could, but flailing a little for the rung with his feet. I wasn't going to try that right away, I decided, so I leaned on my stool, fingertips to the surface of the thing, trying to seem poised. Now the audience was silent.

"Thanks for coming," I murmured to the darkened mass. Then from out of the dimness strode JD. Standing just in front of the pews, she faced the audience and briefly introduced them to the art center. She talked a while and finally turned to us. "Today we have Jim Paul and his friend Harry," she said finally.

Surely, I thought, I had told her Harry's last name—hadn't I?—but she didn't say it, and I could feel him bristling there on my right. Sure enough, he was flared and clenched. It was a stark moment. I remembered what he'd said about his one reason for appearing in public—not for glory, but for balance. "If I'm up there," he'd said, "I won't be just a character." But now he knew he was wrong about that. He wasn't even going to get a last name. He was furious, pretending to be calm, and covered with burrs, besides. A specimen.

Something bad was going to happen, I thought. Harry might explode. I tried not to look at him. If I left him alone with his stage fright, I thought, maybe he'd calm down. But even without seeing him, I could feel the brunt of his not wanting to be there.

JD went on. "They are going to show some slides and tell you about this project, which they did to explore the military history here in the Headlands."

I took a breath and began speaking, taking it away from JD as best as I could, and trying to sound scientific. "We began this project as a process of observation," I said. From the scientific point of view, I didn't want to mention the urges of the Red Creek quartzite, or that we had begun the project on a whim. Or for that matter, that my partner had once shot a TV with a bow and arrow, I thought. I spoke as coolly as possible, about our interest in the technology of conquest. In my peripheral vision was spotlit Harry, glaring. What was I saying? Where was I taking this? I didn't know myself. Words came out of my mouth for a while, and then I waved to Howie. We'd begin with slides, I said. It was Harry, I thought, not the crowd, who made me nervous. The crowd was some dark outline out there.

The pictures indeed behaved differently, revealingly, there in the lecture. I had chosen their order myself, beginning in reverse. Howie had arrived—with the ranger—too late on the day of the shoot, so he hadn't taken any pictures of the setup. To fill this gap in the documentation, I had simply arranged the pictures of the catapult's disassembly in reverse, so that we would appear to be putting it up. The picture of us bolting on the trigger would actually be a picture of us unbolting it. That would make a good introduction, I thought. It looked the same, anyway, in stills, and nobody would know. Only Harry would know, rather, and he wouldn't be a problem. Or so I'd thought. I hadn't even told him about reversing the pictures. What he didn't know wouldn't hurt him, I'd thought.

So the slides went fine, until Howie projected a close-up of Harry and me at work. I said, "Here we are stringing the bow."

"We're unstringing it, actually," Harry said. A cruel pause ensued.

"This is the introduction," I said to Harry.

"I don't care what it is," Harry said. "We're unstringing the bow."

I looked beseechingly at my partner. He looked impassively back. He was mad. "Next slide," I said. I threatened Harry, without moving my lips. "You want it, I'll give it to you," I said.

But after that, I hardly dared to narrate the slides. I said only the most innocuous things and put all of us in la-la land. So the pictures went by, at the stately pace I had told Howie to run them, at thirty seconds each. Granite. Harry pulling the rope. More granite. Me pulling the rope. One picture of the stone in flight, nearly indistinguishable from the flecks on the screen, a blur in the blank sky, and the ocean, supine in the background, seeming to be the subject of the photograph.

But even with my innocuous narration, the pictures were unsatisfactory. For one thing, I looked too stocky, I thought, lumbering around the catapult in my big gloves and mugging for the camera, as Harry in his ILM T-shirt attended to the machine and looked serious. Maybe I was the Igor in the piece, I thought. Howie's action shot from the depths of the bunker made me look particularly goony, my stalking figure back-lighted and thick. Thirty seconds per picture seemed like a rash and ignorant decision by then. Ten seconds, I thought, would still be too long.

"You can see we were into it," I chirped.

For the last picture, I had chosen a humorous shot of myself, which I thought would make a lighthearted end to the slides. In it I appeared to concentrate as I wrote the word Mona on the catapult's white stand. Nobody laughed, though, and the picture began to look more and more stupid. At one point it became stupid enough that I couldn't bear witnessing it any longer.

"Here we are christening the catapult, well, I guess that's it for the slides." I said.

A woman's voice piped up from the front. "What's that you're writing?" she said.

"Mona," I said. "We named it."

At my back, Harry spoke up. "You named it," he said.

We all considered the picture. The image of me grinning with a ballpoint pen and the word Mona distinctly on the white four-by-four of the catapult stand went by, tick by tick. When the image finally disappeared, the screen was mercifully dark for a second, and then, at the back JD turned on the house-lights. There the audience appeared, suddenly, in David Ireland's room. The dads sat in a cluster with their sons. The cowgirls looked on from the front. Our pictures of the catapult were over; now it was our turn to be the objects of attention.

And next to me sat Harry, slumped and blank, looking unrepentant for exposing me and my stupid Mona. All right then, Harry, I thought, you can just take it away. I addressed the audience in a polite tone. "Now Harry's going to tell you something about the history of the catapult," I said. Harry bolted upright a little.

He gave it a good try, he did. But the surprise introduction didn't help. And what he said somehow began with the name Aristotle, by which, a few fatal seconds later, he seemed to realize he'd meant Archimedes. But by then he couldn't stop for the name, and he clearly spent whatever rage he possessed in the effort of thrashing ahead into Rome, which he ruined for a couple of sentences before he knew that the game was up. Then the history dragged to a halt, and he succumbed in disgust, just muttering "God damn it," and falling silent.

For a while I just looked at him, burning him with the focus, as if waiting to see if he had any more to say. But Harry just sat there beneath the catapult, mute and bedraggled, his shoes clotted, burrs on his back. "Maybe I should just sum up," I said at last.

In the audience, Melissa's hand shot up, followed after a second by the fat kid's. I called on the fat kid.

"How far would you say you shot that rock, anyway?" he said. I couldn't tell if the question was a put-up job from his fat dad, and tried to answer in the cheerful and vague way one sometimes speaks to children.

"Oh, a long way," I said.

"How far exactly?" said the kid.

"A couple hundred yards, anyway. Wouldn't you say that, Harry? A couple hundred yards, something like that?"

Disgusted Harry said, "A hundred yards," adding for good measure, "Tops."

"Did you hit the water?" said the kid.

Melissa still had her hand up, so I called on her. "We want to thank you guys for coming here today," she said. "It was entertaining, to say the least. But, really, why'd you name it Mona?"

Mona I thought I'd left behind. I heard one of the dads say, more or less under his breath, "For crying out loud."

Harry was looking at me, of course, smug in the truth of his assertion. I *had* named the catapult, and no one else could answer the question. When I did it, I had been thinking about Ol' Betsy, actually, one of the guns that won the West, but whose gun she was—Davy Crockett's? Daniel Boone's?—I couldn't remember, only that whose-ever gun Ol' Betsy was, she was just a Disney Production from my television childhood, as far as I knew. Not an authoritative source, I thought, and so in the silence murmured, "Just a joke."

"Doesn't look like any kind of female thing that we've ever seen, that's all," said Melissa, as the cowgirls giggled.

"I don't know," I said. "Mona—maybe after the Mona Lisa, whatever." Then there was another hand up, thank God, and a friend's, too: Howie's. I called on him.

"You said you began this project as a process of observation. How did you guys see yourselves when you did this?" Jesus, I thought. Leave it to a photographer.

We had observed things, I thought. But nothing came to mind. "We tried to observe the process as much as possible," I said.

One of the cowgirls piped up. "How much was possible?" she said.

Harry the burr-ball next to me would be no help. Fun, I wanted to say. It was fun. All right, we had fun. "Look," I said. "All I can say is that we tried to observe the process. You saw the slides. We told you how we did it."

"Come on, you guys," said Melissa, teasing us. "Aren't you

owning up to anything up there? I mean you're under the thing—you're sitting up there under the thing."

Sitting up there under the thing, I didn't answer.

Harry said, "Not me. I don't feel like I'm owning up to anything."

The cowgirl who had piped up before did so again, in a fatuous voice, "You didn't do any thinking about phallic symbols, then, when you made this?" This got a laugh from some people in the audience. The dads, though, had had it. As one they rose and began pulling their sons out of the pew. This was dirty talk for them, I guessed. Still, I welcomed their commotion, and took advantage of the stir to attempt to adjourn the lecture.

"We'll answer any of your other questions individually," I said. "And you can come up and look at the catapult if you want."

At this a few more people stood up, tripping the dads' commotion into a general reaction as the idea caught on and the lecture actually ended. After a minute or so, Melissa got up and came over, grinning.

"Grilled," I said. "Gimme a break."

"Why do boys always think that girls just want to spoil their fun?" she said, heading off to join the other cowgirls, who were waiting by the door. When they left, Harry and I only had to answer a few innocent questions. After that, Sara departed with her girlfriend; Howie packed up and said goodbye, and JD came over to check on us, to make sure we weren't planning on leaving the catapult set up in there. She was in a good mood, by then. Mock rocks hadn't come up. "It went better than I thought it would," she called from the doors.

Then Harry and I were left by ourselves in the big room, and we began to take down our catapult. We unstrung the bow, unbolted and removed the springs. We knew how, by then, and didn't have to say anything as we worked. We hauled the parts back down into the basement and left them on the scrap parts heap. By then we weren't sentimental. We put the tools back in the truck and drove out of there, not speaking until we had left the Headlands and picked up the freeway back across the Golden Gate. Then we began listing the things we should have said.

Harry thought the cowgirls had exhibited shallow thinking. He hadn't liked being asked if he was owning up to anything. "All I'm trying to say," he said, "is that if you're not willing to observe this, if you're just going to condemn it, you're never going to see it. It's there, whether you like it or not."

Of course I agreed. "And we did observe it," I said. "But we didn't think much about it, Harry—you've got to admit that. We had fun."

"First of all," said Harry, as we passed beneath the north tower of the red bridge, "you had fun. For me, it was just a bad deal all around.

"And second of all," he said, "as far as thinking about it— that's what you never understood about this whole thing."

"What?" I said.

"This is stupid human life-force we're dealing with here," he said. "Think how old it is, Jim. Think how huge it is. Understanding it doesn't make any difference."

He was right. That hadn't occurred to me. I had assumed that we had failed in not coming to some understanding about the catapult—that it was good, bad, whatever. Harry was saying that the problem was deeper than that. That even if we had understood it, it wouldn't have mattered. Judgment didn't matter. It was what we were, not what we thought.

CHAPTER 27

The Flotsam House

At home on the bowlegged table in my front hall sat the pink-and-white lump of Red Creek quartzite, big as a grapefruit, heavier than a telephone. Now all of its ancientness had a new glimmer that said, "Understanding it makes no difference." It confirmed my first thought, that holding an old rock *was* like looking at the stars, but not in the way that I had imagined—not in oneness with the immensities, but in service to them. The Big Bang was a kind of catapult, it said, and light itself a kind of bombardment. In our heart of hearts, we besiege—and find it fun. And in this we were as stupid, as specialized as dinosaurs.

But what the stone and Harry insisted didn't stick with me somehow, though I couldn't argue with it. That couldn't be it, was all I thought. I had a habit of believing in judgment, maybe. You have a habit of believing in authority, Harry said.

Harry and I remained friends. Afterward I dogged him to admit that he had had fun, and at first he wouldn't give in on the point. It made him shudder to think about our lecture. "Catapult"—he would spit the word out—"All I got out of that was that I could make something. If you can make something you can make anything." He griped for a while, but after that his bitterness about the catapult moderated, and the whole business began to seem funny, it was so stupid.

About a month after the catapult, I moved to Boston on a temporary assignment. Leaving the Red Creek quartzite and Harry behind wouldn't help, though. Alone in an attic room in snowy Massachusetts, I read *War and Peace*, and in the year after the catapult project—traveling on assignment more often than not—I continued to study catapults and sieges and to sightsee on the subject. I went to Meteor Crater, Stirling Castle, Jerusalem. Maybe I'd come up with something, I thought, if I knew the whole story. Maybe I could refrain, if I knew from what.

But before I left San Francisco, in that intervening month, Harry and I goofed around with the kids again on the weekends—all three kids, since Julia was now getting big enough to take along with the boys. We owed his family some time, we figured. And on the weekend before I left for Boston, at the end of the year, while Susan and Sara went to brunch, Harry and I took the kids out to the beach in the Headlands. I recalled our catapult parts, lying on the scrap heap up there in the art center. The stand had already been cannibalized for building maintenance, I'd heard. "Maybe we could retrieve some of the crucial stuff," I suggested to Harry.

But he'd have none of it. "That was a bad deal all around," he said again.

"Just for camping trips?" I said.

"No way," he said.

Sometimes on the beach Harry and I would play a game called Postnuclear Survival. We'd play it when we got tired of collecting bones and shells, by combing the flotsam on the beach for things that might prove useful in the work of starting culture over. A piece of driftwood shaped like a hoe was a good find then, and fishing line, rusted knives, Styrofoam, all excellent. Once, as part of this game, Harry made arrowheads

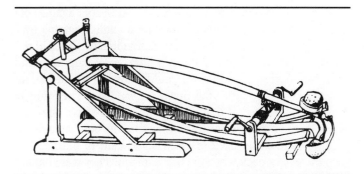

out of bits of sea glass, saying, "I'll be a valued member of society someday."

It wasn't a game we played with the kids, but that day the sea had cast up loads of flotsam. Winter storms had finally arrived, one of them earlier in the week knocking apart decks and piers in Mendocino. So all this stuff, redwood mostly, had come south with the swell, and a lot of it was thrown up there in the southern crook of Rodeo Beach, where we were.

"Beams," I said, when I saw the piles of driftwood. "Maybe we could make another catapult."

The boys chimed "Yeah!" But Harry just said no.

"Even though," he added, "we could make a really good one this time."

"We could make it Postnuclear," I said, gesturing at the heaps of wreckage.

"Building a catapult then would be making the whole mistake all over again," said Harry.

"What do you mean, Papa?" Isaac chimed up. "We'd have to defend ourselves."

"You're getting too smart for your own good," Harry said to his son.

In the end we decided to build a flotsam house, which was

a lot of work. Harry and I leaned two big beams against the sea cliff. The boys and Julia helped for a while, bringing over smaller pieces of wood, which we used to line the sides of the lean-to. Harry and I roofed the gap, then shored up the posts with the heaviest stuff that we could move on the beach, so that the house wouldn't fall down on the kids. They mostly couldn't have cared at that point, ignoring us as we began to elaborate the structure by adding a porch. The boys went down to the surf line and tried to build some kind of blockade to keep the waves back when the tide came in. They were bailing in no time. Julia sat in the hut and happily dug holes. We'd sit in the house for a while, then decide it needed something else and set to work again. We ended up working all day, having fun. When we left, we gave the house away to some other people who'd showed up.

About the Author

JIM PAUL's poems have appeared in *The New Yorker, The Paris Review, Poetry, The Quarterly*, and elsewhere. He was a Hopwood Award winner at the University of Michigan and a Wallace Stegner Fellow in Creative Writing at Stanford University.